Dictionary of Geotechnics

Dictionary of Geotechnics

S. H. Somerville, CEng, FICE, FGS

M. A. Paul, BSc, PhD, FGS

Butterworths

London Boston Durban Singapore Sydney Toronto Wellington

First published 1983

© Butterworth & Co. (Publishers) Ltd, 1983

British Library Cataloguing in Publication Data

Somerville, S. H.
 Dictionary of geotechnics.
 1. Geotechnics — Dictionaries
 I. Title II. Paul, M. A.
 624′.03′21 QE601

 ISBN 0-408-00437-1

Printed and Bound by Robert Hartnoll Ltd, Bodmin, Cornwall

Photoset by Mid-County Press, London SW15

Preface

The aim of the authors has been to provide a dictionary of geotechnical terms for use by practising engineers, advanced students and those who by the nature of their professions and interests, from time to time require basic information about the subject.

The book includes terms in everyday use by geotechnical engineers and covers the fields of soil and rock mechanics, soil and rock engineering, site investigation, hydrology, and the various geotechnical practices such as dewatering, ground stabilisation, earthmoving and compaction, trenching and excavation. A selection of terms is included in the subjects of engineering geology and engineering geomorphology while practical information with typical values and results are tabulated to make it a useful reference work.

Geotechnics is a relatively young and developing subject and the authors realise that the list is not exhaustive. Reference material has therefore been included to point the reader towards additional study where necessary, rather than to include a greater number of shorter entries.

Grateful acknowledgement is made to the British Standards Institution for permission to reproduce definitions given in various BSI publications, to the American Society of Civil Engineers for permission to reproduce definitions concerning the phenomena of liquefaction prepared by their Committee on Soil Dynamics, and to the numerous authors of books and technical publications to which the authors have made reference at some time during their respective careers.

S.H.S.
M.A.P.

Abbreviations of authorities

AASHO	The American Association of State Highway Officials
API	American Petroleum Institute
ASCE	American Society of Civil Engineers
ASTM	American Society for Testing and Materials
AWWA	American Water Works Association
BPR	Bureau of Public Roads
BRE	Building Research Establishment
BS	British Standard
BSI	British Standards Institution
BSCS	British Soil Classification System
C&CA	Cement and Concrete Association
CIRIA	Construction Industry Research and Information Association.
CP	Code of Practice
DOE	Department of the Environment
FAA	Federal Aviation Agency
HMSO	Her Majesty's Stationery Office
ICE (Inst. CE)	Institution of Civil Engineers (UK)
ISSMFE	International Society of Soil Mechanics and Foundation Engineers
ISSS	International Society of Soil Sciences
MEXE	Military Engineering Experimental Establishment
MIT	Massachusetts Institute of Technology
NGI	Norwegian Geotechnical Institute
NHBC	National House Building Council
PCA	Portland Cement Association
TRRL	Transport and Road Research Laboratory
UOP	Universal Oil Products
USBR	US Bureau of Reclamation
USDA	United States Department of Agriculture
WES	Waterways Experiment Station

Related British Standards Institution publications

CP2 : 1951	Code of practice for earth retaining structures
CP101 : 1972	Foundations and substructures for non-industrial buildings of not more than four storeys
CP110 : 1972	The structural use of concrete
CP1013 : 1965	Earthing
CP1021 : 1979	Cathodic protection
CP2004 : 1972	Foundations
BS1377 : 1975	Methods of test for soils for civil engineering purposes
BS4019 : 1974	Core drilling equipment
BS5930 : 1981	Code of practice for site investigations (superseded CP2001 : 1957)
BS6031 : 1981	Earthworks (superseded CP2003 : 1959)
PD6031 : 1978	The use of the metric system in the construction industry
PD5686 : 1972	The use of SI units

Short list of publications

'A guide to the structural design of pavements for new roads', Road Note 29, 3rd edn, Department of the Environment, UK Transport and Road Research Laboratory, HMSO, 1970.

Dictionary of Geological Terms, Prepared Under the Direction of the American Geological Institute, Anchor Books, Anchor Press—Doubleday, New York, 1976.

Earth Manual, US Department of the Interior, Bureau of Reclamation, US Government Printing Office, tentative edn 1951, 1st edn 1963, 2nd edn 1974.

Encyclopedic Dictionary of Exploration Geophysics, R. E. Sheriff, Society of Exploration Geophysics, 1973.

European Symposium on Penetration Testing, Stockholm, 1974.

Foundation Design and Construction, M. J. Tomlinson, Pitman, 1969.

Foundation Engineering, R. B. Peck, W. E. Hanson and T. H. Thornburn, Wiley, New York, 1973.

Foundation Engineering Handbook, Hans F. Winterkorn and Hsai-Yang Fang, Van Nostrand Reinhold, 1975.

Foundation Instrumentation, T. H. Hanna, Trans Tech Publications, 1973.

Fundamentals of Soil Mechanics, D. W. Taylor, Wiley, New York, 1948.

Groundwater Hydrology, Herman Bouwer, McGraw-Hill, 1978.

International Society for Rock Mechanics. Commission on Standardisation of Laboratory and Field Tests, *International Journal of Rock Mechanics — Mining Sciences and Geomechanical Abstracts*, **15**, 319—368, 1977.

Introduction to Earthquake Engineering, Shunzo Okamoto, University of Tokyo Press, 1973.

Methods of Treatment of Unstable Ground, F. G. Bell, Newnes-Butterworth, 1975.

Pile Design and Construction Practice, M. J. Tomlinson, Viewpoint Publication, Garden City Press, 1977.

Recent Developments in the Design and Construction of Piles, The Institution of Civil Engineers, London, 1980.

Rock Slope Engineering, E. Hock and J. W. Bray, Institution of Mining and Metallurgy, London, 1977.

Soil Mechanics in Engineering Practice, K. Terzaghi and R. B. Peck, Wiley, New York, 1948.

Specification for Road and Bridge Works, 5th edn, Ministry of Transport, HMSO, London, 1976.

'State-of-the-art review of soil sampling in the United Kingdom', K. W. Cole, April 1979.

Theoretical Soil Mechanics, K. Terzaghi, Wiley, New York, 1943.
Water Well Technology, M. D. Campbell and J. H. Lehr, McGraw-Hill, 1973.

A

AASHO   The American Association of State Highway Officials.

AASHO density tests   *See standard Proctor compaction test.*

AASHO soil classification system   This system evolved from the US Bureau of Public Roads system of classifying soils in accordance with their performance as sub-grades beneath road pavements. It is divided into seven basic groups, A-1 to A-7, with sub-divisions in Groups A-1, A-2 and A-7, the overall quality generally decreasing with increasing classification. The system adopts classification by sieve analysis and index property tests, and differentiation between the quality within a particular group is made by the group index (GI):

$$GI = (F - 35)[0.2 + 0.005(LL - 40)] + 0.01(F - 15)(PI - 10)$$

where F = the percentage of fines passing No. 200 sieve; LL = the *liquid limit* (%); and PI = the *plasticity index* (%). The group index is written (in parentheses) following the soil group and indicates the general quality of a soil as a sub-grade material:

excellent	A-1a (0)
good	(0–1)
fair	(2–4)
poor	(5–9)
very poor	(10–20)

References: *Standard Specification for Highway Materials and Methods of Sampling and Testing Part 1*, AASHO, 1970; *Soil—Cement Laboratory Handbook*, Portland Cement Association, 1971.

ablation   Loss of material, particularly used to refer to the loss of ice from a glacier by any process—e.g. calving, melting, etc.

Abney level, or clinometer   A small hand-held optical instrument for measuring vertical angles, comprising a hand level with an attached index arm and graduated arc. In use, a milled wheel is rotated until the reflected image of a bubble appears in the line of sight, when the required angle is read on the arc.

abrasion   The loss of material from one or both of two surfaces that are in contact and undergoing relative motion. In detail, abrasion is the result of the shearing off of interlocked projections from the two surfaces.

abrasion test   *See Los Angeles abrasion test; sand blast test.*

absolute temperature   The pressure of a gas reduces linearly with decreasing temperature and when zero, has an apparently unique temperature of about $-273.15°$ Celsius. This temperature is termed *absolute zero* ($0°$ Kelvin) and is a convenient reference point for all other temperatures which are then known as *absolute temperatures*.
   Reference: *Introduction to Physics for Scientists and Engineers*, F. J. Bueche, McGraw-Hill Kogakusha Ltd, 1975.

absorption   An electrostatic or chemical process whereby molecules or ions penetrate into the interior of a solid or liquid. *See electromagnetic absorption.*

absorption spectroscope   An instrument used to measure the amount of energy absorbed by a particular type of material or surface, by spectral analysis techniques.

abyssal deposits   Sediments that occur in the deep ocean basins. They fall into one of three classes: (a) oozes, formed from the shells of planktonic organisms and which may be either calcareous or siliceous; (b) red clays, formed from volcanic dust; (c) detrital sediments, introduced from the continental margins via turbidity currents.

Abyssinian tube well   An early type of wellpoint, consisting of a pointed, perforated tube that was driven into the ground and from which water could be subsequently pumped.

accessory minerals   Those minerals which may be present or absent in a crystalline rock without prejudice to the identification of that rock, and are thus contrasted with the essential minerals, on which the identification is based.

acetic acid   An acid used for stimulation and reconditioning of water wells and for overcoming a variety of problems in their completion.

acid dip survey   A method of measuring the inclination of a borehole by placing a bottle of acid at the required depth for a short period and subsequently measuring the angle of etching formed around the inside of the bottle.

acidising   The process of introducing acid into the ground (in a *well*) in order to make it more permeable by dissolving acid-soluble material present in the pores and also to remove encrustation from *well screens* and *gravel packs* by dissolving cementitious material.

acidity   An excess of hydrogen ions over hydroxyl ions which in groundwater is usually caused by uncombined carbon dioxide, mineral acids, and salts of strong acids and weak bases such as iron

2

and aluminium salts from mine waters or industrial wastes. *See pH value*.

acid rock   An igneous rock in which quartz appears as an essential mineral; hence, a rock which had a sufficient proportion of original silica to satisfy the requirements of the metal oxides during crystallisation.

acoustic   Notionally, to do with sound waves. Used in marine geophysics to refer to energy sources that produce considerable energy in the human audio range (e.g. *boomers* and *pingers*). Also used in *rock mechanics* to refer to the transmission of high-frequency sound waves in testing and strain measurement.

acoustic blanking   The dispersion and absorption of seismic energy by sediments which contain gas bubbles in the pore spaces.

acoustic strain gauge   An instrument for measuring the extension or contraction of a structural member incorporating taut wires which, when plucked, emit a note depending on the tension in the wire.

acoustic-velocity log   *See sonic log*.

action factor   The ratio of the indicated pressure to the true pressure recorded by an *earth pressure cell*.

Reference: *CIRIA Technical Note 96*, February 1979, I. K. Nixon and B. O. Skipp.

active earth pressure   The value of lateral pressure exerted by a soil on a retaining structure that is reached when the structure yields and allows the soil to expand in a horizontal direction. The *coefficient of active earth pressure*, K_a, is equal to $(1-\sin\phi)/(1+\sin\phi)$ where ϕ is the *angle of internal friction* of the soil.

active layer   That part of the *permafrost* which, being at surface, is subject to seasonal episodes of thawing and freezing. The thickness of the active layer varies with location, climate and soil type, but is, in general, of the order of a few metres.

active pressure   *See active earth pressure*.

activity of soil   The ratio of the *plasticity index* of a soil to its clay fraction. Shrinkage and swelling of soils due to changes in moisture content are a function of soil activity, and the application of Casagrande's plasticity chart *(Figure A.1)* and Skempton's activity chart *(Figure A.2)* provides useful indications of possible problems in this respect. *See also soil description and soil classification*. Typical values of activity for various clay minerals are:

Clay mineral	Activity
Muscovite	0.23
Kaolinite	0.40
Illite	0.90
Ca-montmorillonite	1.50
Na-montmorillonite	6.00

3

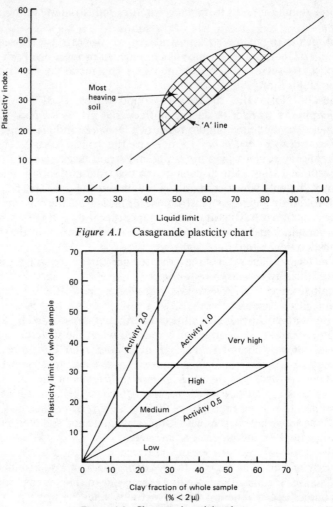

Figure A.1 Casagrande plasticity chart

Figure A.2 Skempton's activity chart

adfreezing The adhesion of *frost-susceptible soil* in contact with foundation walls due to freezing resulting in sufficient bond to transfer heaving pressures from the soil to the structure. Also referred to as *frost-grip*.

adhesion *See skin friction (piling)*.

adit An approximately horizontal tunnel driven to underground workings.

adjoining owner Adjoining owner is defined for the purposes of demolition, shoring and underpinning works as a **person or**

4

persons owning land, a building or portion of a building which abuts upon or is within 3 m of the proposed building operations. Where the excavation is for a new building, and particularly where underpinning has to be done, buildings up to a distance of 6 m should be included.

Reference: CP 2004: 1972.

Admiralty Tide Tables   Tables giving daily predictions of times and heights of high and low waters at a selected number of 'standard' ports. Usually based on continuous observations of the tide over a period of at least one year at that standard port. Tables also list data for 'secondary ports' to enable predictions to be calculated on the basis of the predictions given for the standard ports.

adobe   An American term for a heavy-textured soil, having a high percentage of colloidal clay, which tends to break into irregular, roughly cubical blocks on drying.

adsorption   Condensation of a gas or liquid on the surface of a solid.

aeolian   Of, or pertaining to, the wind. Usually applied to sediments that have accumulated from wind-borne particles.

aerobic   Living or active only in the presence of oxygen.

affine   In structural geology, used to describe a style of deformation in which originally parallel elements remain parallel although circles are deformed to ellipses. Also known as *homogeneous deformation.*

aggregate   The constituents, comprising *sand* (fine aggregate) and *gravel* or crushed stone (coarse aggregate), used in the manufacture of concrete or bituminous surfacings.

aggregate abrasion value test   *See Dorry test.*

'A' horizon   The upper part of the soil profile, which extends to about 2 m below ground level and whose physical properties are subject to change by seasonal moisture content and temperature variations and by biological agents such as roots, worms and bacteria. The 'A' horizon is subject primarily to mechanical effects of weathering and leaching. *See 'B' horizon; 'C' horizon.*

airborne magnetometer   An instrument carried aboard an aircraft for measuring the Earth's magnetic field.

air-dried   Allowed to dry without the application of artificial heating. Air-dried soil usually has at least about 5 per cent moisture.

air-entry permeameter   A device for measuring the *permeability* of relatively dry soils. It comprises a metal cylinder of about 300 mm diameter which is driven 100–200 mm into the surface of the soil and to which a lid assembly with standpipe and reservoir can be clamped.

Reference: *Groundwater Hydrology*, H. Bouwer, McGraw-Hill, 1978.

5

air-entry value The pressure of air that must be applied to the surface of a porous material in order to force air to enter the pore space if it is initially saturated with water. The air-entry value is used to indicate the coarseness of porous stones, etc.

airfield classification systems Several such systems were developed in the early 1940s to categorise soils relevant to airfield pavement sub-grades. Until then, highway pavement design was not applicable to most military airfields being constructed at that time. These systems included the Civil Aeronautics Administration (CAA) System and the Airfield Classification (AC) System, and modification of the latter led to various other systems, including the *unified soil classification system.*

airfield cone penetrometer An instrument, developed by the Waterways Experiment Station of the US Army Corps of Engineers, used to measure soil strength and its suitability for potential aircraft landing. Similar to that used in the *WES mobility cone penetrometer test* but with a smaller base area and thus suitable for measuring stronger soils. The force to push the penetrometer into the ground is called the *Airfield Index (AI)*. In frictional soil the rate of increase of cone penetration resistance with depth is termed the *Cone Index Gradient (CIG)* and gives a measure of soil density. Sand numbers (N_s) and clay numbers (N_c) have been derived giving pull and torque coefficients for tyre performance.

Airfield Index (AI) *See airfield cone penetrometer.*

air flushing The process of using compressed air to cool the drill bit down a borehole and to lift the cuttings from the hole.

air guns Instruments used in marine geophysical surveying that allow a continuous flow of compressed air bubbles into the water behind the survey vessel. The oscillation of the bubbles as they alternately expand and contract generates a sonic wave, the reflections of which from the sea-bed and sub-bottom strata are recorded and can be interpreted to determine depth and type of the stratification. *See continuous seismic reflection profiling.*

air lift Apparatus used for raising water in a well or removing sediment from below water level, comprising a rising main (eductor pipe) into the base of which compressed air is fed to aerate the water. This lowers the specific gravity of the water in the pipe relative to the surrounding water and causes it to rise up the pipe, taking any sediment with it. The efficiency of the system depends on the relationship between submergence and required lift, size of rising main, and quantity and pressure of air supplied.

air-lift pump *See air line.*

air line (1) In an air-lift pump, the smaller vertical air line usually submerged to within a short distance of the base of the eductor pipe, the depth below water level of which is used in calculating the air pressure required to start the air lift. (2) In diving, the air supply tube between the diver and the surface demand air pump.

air lock In excavation works for tunnels, cofferdams and caissons where air is pressurised to exclude groundwater and stabilise the formation, the air lock comprises a chamber whereby men and materials may pass from the pressurised zone to atmospheric conditions outside.

air photographs Used in the preparation and revision of maps and plans, and to assist in the identification of geological and geomorphological features and the interpretation of previous uses of a site.

air void ratio The ratio between the volume of voids occupied by air in a soil mass and the volume of solids.

air voids content The proportion of air in a soil mass. In general, when soil is compacted, it is not possible to expel all the air within practical limits of compaction and about 5–10 per cent of air is retained. *(See Figure A.3)*.

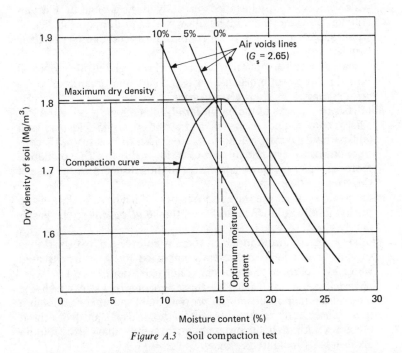

Figure A.3 Soil compaction test

air voids line A line showing the dry density/moisture content relationship for soil containing a constant percentage of air voids. (Air voids lines are shown in *Figure A.3*.) The line can be calculated

7

from the equation

$$\rho_d = \rho_w \frac{1 - \dfrac{V_a}{100}}{\dfrac{1}{G_s} + \dfrac{w}{100}}$$

where ρ_d = the dry density of the soil (Mg/m^3); ρ_w = the density of water (Mg/m^3); V_a = the volume of air voids in the soil, expressed as a percentage of the total volume of the soil; G_s = the specific gravity of the soil particles; and w = the moisture content, expressed as a percentage of the mass of dry soil.
    Reference: BS 1377: 1975.

ALGOL   The computer language acronym for ALGOrithmic Language.

alidrain   A band-shaped drain strip comprising a cellulose filter which can be rapidly installed vertically in the ground by a 'drain-stitcher' machine. Such filters allow water to flow out of the soil when subjected to a surcharge and thus accelerate settlements (similar to *geodrain*).

'A' line   A line shown on a *plasticity chart* giving an empirical boundary between inorganic clays and silty and organic soils. *See soil description and soil classification.*

alkali basalt   Rocks of the basalt family which have a generally low silica content, and in which sodic and potassic feldspars and feldspathoids occur. Such rocks are often associated with continental rift valleys, and appear to be formed by melting of mantle material under pressures equivalent to those found near the base of the continental crust.

alkali gabbro   Rocks of the gabbro family which have a relationship to that family analogous to that of the *alkali basalts* to the basalt family.

alkalinity   A condition in which there is an excess of hydroxyl ions over hydrogen ions, and usually imparted to natural or treated water by bicarbonate, carbonate and hydroxides.

alkali rock   A term applied to those igneous rocks in which the minerals contain a substantial proportion of sodium or potassium ions. These ions commonly enter the feldspar or felspathoid minerals, although ferromagnesian minerals may also contain alkali metals in some instances.

Allen–Hazen formula   *See effective grain size; permeability.*

allochthonous   Out of place or transported. Used particularly in the cases of (a) sediments that have been transported into an area in which they are clearly foreign; (b) tectonic transport of identifiable beds by overthrust faulting, particularly the nappes of the European Alps. *See also* **autochthonous**.

allowable bearing pressure *Allowable net bearing pressure* is the maximum allowable *net loading intensity* at the base of a foundation, taking into account the *safe bearing capacity* of the ground, the amount and kind of *settlement* expected and the ability of the structure to accommodate that settlement. It is therefore a combined function of both the site conditions (including all construction in the vicinity) and the characteristics of the particular structure it is proposed to erect. *Allowable gross bearing pressure* is the same as allowable net bearing pressure, except that it includes the **overburden pressure** above the level of the foundation base. The latter term is not often used in practice.

allowable gross bearing pressure *See allowable bearing pressure.*

allowable load In *piling*, the safe load after taking into account all relevant factors such as its *ultimate bearing capacity*, pile spacing, overall *bearing capacity* of the ground below the piles, allowable settlement and possible drag-down due to any surface compressible deposits.

allowable net bearing pressure *See allowable bearing pressure.*

alluvial fans Debris fans or cones that are built out into a main valley from a side valley by the action of strong periodic run-off. Despite the name, the debris is usually poorly sorted, with a coarse component in the upper parts of the fan. Transport of material to and across the fan is often by *debris flow*, in which many sizes of particle are carried. Alluvial fans are characteristic of poorly vegetated regions that suffer from periodic flash floods.

alluvial plain A phrase used in general terms to describe the flat-lying zone adjacent to the channel in the lower parts of a river system, which is often subjected to periodic flooding. In more precise usage it is the surface of the alluvial valley fill that has been deposited by the river.

alluvium An imprecise term used to describe sediments (usually fine-grained) that are transported and deposited in the lower reaches of a river system.

alpha (α) method of pile design A total stress method used to determine average unit skin friction (f) between mainly cohesive soil and pile shaft in a soil layer of thickness (i) whereby $f = \alpha C_u$, for which α depends on the undrained shear strength (C_u) of the soil and varies between about 0.2 and 1.2 (decreasing with increasing C_u). *See also* **beta (β) method** and **lambda (λ) method of pile design**.
    Reference: API *Recommended Practice for Planning, Designing and Constructing Fixed Offshore Platforms*, Report API RP 2A, 8th edn, March 1977.

alpha pile A driven and cast-in-place concrete pile at one time installed by Balkan Piling Ltd *(Figure A.4)*. An outer casing and an internal hollow mandrel are driven simultaneously, the casing having a detachable cast-iron shoe filled with concrete in which steel anchor loops are embedded to secure vertical reinforcement rods. By progressively withdrawing the mandrel, replenishing it with concrete and redriving, a dense pile is formed with no contamination of the concrete, waisting of the shaft or segregation of the mix.

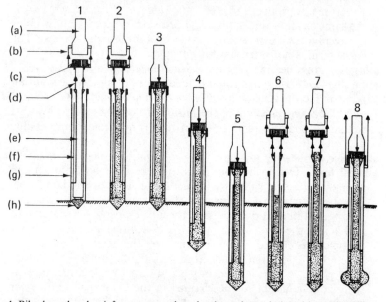

1 Pile shoe placed, reinforcement anchored, tube and mandrel positioned for filling (a) Hammer, (b) helmet suspension slings, (c) helmet and dolly assembly, (d) mandrel suspension slings, (e) mandrel, (f) reinforcing bars, (g) outer tube, (h) cast-iron pile shoe
2 Mandrel and tip of tube filled with concrete
3 Helmet and hammer lowered
4 Tube and mandrel being driven
5 Driving completed
6 Mandrel raised (some concrete has passed from mandrel to tube)
7 Mandrel replenished with concrete
8 Tube raised and mandrel driven down simultaneously. Concrete forced out to form bulb

Figure A.4 Preliminary stages in the formation of an Alpha *in situ* pile

altiplanation A high-altitude process supposedly capable of producing level surfaces in mountainous areas. Such surfaces are often strewn with rock fragments and are termed blockfields. The process

10

involves the freeze–thaw mechanism, but the details remain obscure.

aluminium The third most abundant element in the Earth's crust; found in minerals, rocks and soil. It is present in most natural waters as a soluble salt, colloid or insoluble compound.

AM9 A chemical resin used in *grouting* as an accelerator for and to facilitate penetration of fine-grained soils. Polymerised AM9 is an elastic resin whose deformation is proportional to the load applied and has no hysteresis effect. It is useful for filling *fissures* in rock which are liable to movement due to temperature changes.

amortisation The process of clearing off, liquidating or otherwise extinguishing a debt, generally by a sinking fund.

anaerobic The ability to live or be active in the absence of free oxygen.

anastomosing Literally, 'rejoining'. A term used to describe a particular type of river system in which channels divide and rejoin in a complex pattern. Otherwise known as a *braided river pattern*. Such systems are characteristic of proglacial areas, but are also found in other environments where river discharges are high but irregular and the rivers are well supplied with sediment in which easily eroded channels may be cut.

anchor pile *See tension pile.*

andesite An extrusive igneous rock belonging to the calc–alkali association. Characterised by the presence of andesine (plagioclase feldspar with 50–70 per cent sodic ions) in addition to mafic minerals, notably pyroxenes and amphiboles. Quartz is an accessory mineral. Andesites and related rocks are characteristic of volcanism at continental margins; hence the name (derived from the Andes). In plate tectonic terms they are erupted at subduction zones where ocean crust is destroyed by melting at depths below about 50 km.

Andina cone/friction sleeve *See penetrometer (apparatus).*

anemometers Instruments for measuring wind direction and velocity. Types include *cup and vane, hot wire, impeller, laser–Doppler, pressure, sonic, vortex velocity.*

aneroid barometer *See barometer.*

angle dozer A tracked or wheeled tractor provided with a mould board set at an angle to the longitudinal axis of the vehicle.

angle of internal friction A term synonymous with *angle of shearing resistance* but normally limited to cohesionless soils, and equal to the angle ϕ in the Coulomb equation for shear strength, where $S = C + P \tan \phi$ in terms of *total stress*, in which C = the soil cohesion and P = the normal pressure on the potential surface of sliding or shear plane. Modified to ϕ' where $S = C' + (\sigma_n - u) \tan \phi'$ measured in terms of *effective stress. See shear strength.*

11

angle of obliquity   In sliding friction between two bodies, the angle between the resultant of the normal force and applied lateral force and the normal force.

angle of repose   The steepest angle to the horizontal at which a material comes to rest when poured or dumped in a heap or on a slope.

angle of shearing resistance   A term synonymous with *angle of internal friction* but normally limited to strength parameters of cohesive soils. *See shear strength.*

ångstrom   A unit used to express very short wavelengths and equal to 10^{-10} m.

angular distortion   The relative rotation of a point in a structure about another. Limiting values of angular distortion for adjacent foundations have been proposed by various researchers for different structural types. *See Bjerrum's danger limits for distortion of structures* and *Table 13*.

anion   A negatively charged ion or radical. In electrolysis the negative ions move to the *anode*.

anisotropy   Exhibition of different physical properties when measured in different directions. For example, the vertical permeability of a laminated or varved soil mass is often significantly different from its horizontal permeability.

annular space   In well drilling, the space between the well casing and the sides of the borehole.

anode   The *electrode* through which direct current enters an *electrolyte*.

anomaly   In geophysics, a departure from some theoretically expected value of the Earth's gravitational or magnetic field. This departure is assumed to result from the geometric configuration of the materials under investigation, and on this basis may be analysed to give information on their subsurface disposition.

anticline   A geological structure in which the beds are folded so that the centre of the fold contains the oldest rocks. If the sequence has not been inverted, the result is in the form of an arch.

antidune   A sedimentary bed form that is produced in the upper flow regime—i.e. is formed under high discharges. Similar to a dune in appearance, but dynamically it exhibits the opposite behaviour, in that the rapid erosion of sediment and its subsequent deposition on the downstream side causes the dune form to migrate upstream over a period of time.

antifoam   An agent added to acid in well drilling, to prevent or retard foaming during the acid reaction. *See acidising.*

anvil   That part of a power-operated pile hammer which transmits the blow from the ram to the pile.

aperture of discontinuity   One of the ten parameters selected to

12

describe discontinuities in rock masses, being the perpendicular distance between adjacent rock walls of a discontinuity, in which the intervening space is air- or water-filled. *See discontinuity.* Reference: International Society for Rock Mechanics, Commission on Standardisation of Laboratory and Field Tests, 1977.

API American Petroleum Institute. This organisation sponsors standardisation of tools and materials used mainly in the oil industry.

API–CPT method of pile design An adaptation of the *alpha (α) method of pile design* for North Sea conditions where normally consolidated to moderately overconsolidated clays are found to 60–70 m in the Forties Field. For such conditions, the shear strength is taken as 0.2–0.3 times the effective overburden pressure at shallow depth and for greater depths skin friction up to 50 per cent of the measured shear strength is permitted.

API method of pile design A method used in the design of fixed offshore platforms, and similar to the *alpha (α) method of pile design* but wherein upper limits are fixed for the developed adhesion and α, depending on depth below the mudline.

From mudline to 30.5 m:
  developed unit shaft adhesion, c_a = the undrained shear strength, c_u
  but $C_a \ngtr 50$ kN/m^2 and $\alpha = 1.0$ maximum
Below 30.5 m:
  developed unit shaft adhesion $C_a = C_u$ or $C_a = 0.33$ times the effective overburden pressure, p_0', and $\alpha = 0.33$ maximum, whichever is the lower

Considered by Fox to be too conservative for shallow overconsolidated clays and unsafe at depth in lightly or moderately overconsolidated clays having shear strengths equivalent to $0.25-0.65p_0'$. Modification to Krause's method increases C_a at shallow depth. *See also API–CPT method of pile design.*

References: *API Recommended Practice for Planning, Designing and Constructing Fixed Off-shore Platforms*, API RP 2A, 7th edn, January 1976; 'Piling for North Sea installations', D. A. Fox, *Offshore Engineer* offprint 222.1–10, 1975; 'The design of foundation piles of offshore structures', E. R. Krause, Off-shore engineering seminar, Heriot-Watt University, 1973.

API unit A unit of counting rate for radioactivity logging.

apparent angle of internal friction (ϕ_u) A shear strength parameter with respect to total stresses in the equation

$$\text{shear strength } (\tau_f) = C_u + \tan \phi_u$$

where C_u is the apparent cohesion. *See apparent cohesion (C_u).*

13

apparent cohesion (C_u) A shear strength parameter with respect to total stresses in the equation:

$$\text{shear strength } (\tau_f) = C_u + \tan \phi_u$$

where ϕ_u is the apparent angle of internal friction. In an undrained situation with saturated cohesive soils (i.e. $\phi_u = 0$), C_u is also called the *undrained shear strength*. See *apparent angle of internal friction* (*ϕ_u*).

apparent dip The angle made with the horizontal by a bed which is seen in a section that is not parallel to the line of the maximum slope or true dip and will vary down to zero if the section is perpendicular to the true dip.

apparent dry specific gravity The dry weight of a material per unit exterior volume of a material. Used as a *rock mechanics* test to provide an indication of rock strength.

apparent porosity The ratio between the pore space determined from water absorption in a vacuum and the exterior volume of a material. Used as a simple *rock mechanics* test to give an indication of permeability and degree of deterioration due to weathering.

apparent resistivity The resistivity as measured between two electrodes in the ground, and which thus may be affected by several subsurface units. This resistivity will vary as the separation of the electrodes changes, owing to variations in the path taken by the current. The analysis of the pattern of variation induced by changes in spacing is the basis of the resistivity method of electrical prospecting. See *electrical resistivity*.

aquiclude A formation which, although capable of absorbing water slowly, will not transmit it fast enough to produce a significant supply for a spring or well and can be considered as an impermeable *confining bed* relative to adjacent water-bearing strata.

aquifer A soil or rock layer containing sufficient water to allow of the yield of significant quantities to wells and springs. An aquifer may be said to be unconfined if the water table acts as the upper surface of the zone of saturation, or as confined if the groundwater is confined under pressure in excess of atmospheric by overlying relatively impermeable strata.

aquifuge A formation which can neither absorb nor transmit water.

aquitard A rock or soil layer which inhibits groundwater flow.

arching The transfer of a proportion of the overburden pressure to the sides of an opening as a material yields towards the opening (e.g. vertically above a tunnel or horizontally where poling boards are removed from the vertical face of an excavation). The degree of arching increases with an increase in the angularity of the particles, the density of the material and its cohesion.

area of influence The area of land surrounding a pumping well below which the groundwater levels or piezometric water heads are changed during discharge of the well.

area of pumping depression The area of land overlying the *cone of water table depression* caused by *well* pumping.

area ratio The area ratio of a soil sampler is given by the expression

$$\text{area ratio (per cent)} = \frac{OD^2 - ID^2}{ID^2} \times 100$$

where OD and ID are the outside and inside diameters, respectively. Thin-walled tube samples (e.g. *Shelby tube sampler*) have an area ratio of about 10 per cent, while the standard 100 mm open-drive sampler used in the UK has an area ratio of approximately 30 per cent.

arenaceous Of a sediment or sedimentary rock, sandy, composed dominantly of sand-sized particles.

argillaceous Of a sediment or sedimentary rock, clayey, or fine-grained, composed dominantly of clay and silt-sized particles.

ARGO Acronym for Automatic Ranging Grid Overlay, marketed by Cubic Industrial Corporation, USA, and comprising an offshore electronic positioning system capable of an accuracy to within 10 m at ranges up to 740 km (daytime) and 408 km (night-time). The system allows up to eight mobile stations to be simultaneously in operation using two fixed base stations.

arkose A variety of sandstone in which more than about 25 per cent of the grains are composed of feldspar. Arkoses are usually formed from the breakdown of acid igneous rocks, and commonly accumulate subaerially, in which case they often exhibit secondary reddening due to oxidation of ferruginous compounds.

arrival In geophysics, the arrival of energy at a detector or geophone. Used in seismic prospecting; hence *arrival time*, the time elapsed between the release of energy and its detection at a particular position.

arrival time *See arrival.*

arroyo A Spanish term for a small stream or gutter. Usage varies and in some Latin-American countries includes gorges of major proportions.

Artemis A marine positioning system comprising a microwave hydrographic and geodetic position-indicating instrument of the two-station type. Accuracy is within 2 seconds of arc in azimuth and 1.5 m at a distance of 30 km.

artesian Of *groundwater* or other confined fluid, under a pressure greater than that corresponding to the depth of burial—i.e. greater than hydrostatic. Thus, the fluid will flow from an artesian well without pumping. Specifically, the piezometric level for the fluid is above the local ground surface.

15

artesian flow *See equilibrium well formulae.*

artesian well An artesian well is one deriving water from an artesian or confined water body which when penetrated by the well causes water to rise up the well to above the top of the water body.

artificial recharge A method of inducing filtration of water into the soil by pumping into injection wells or increasing the water volume in natural basins, so as to limit the decline of the groundwater level—or raise it. It is used in conditions where it is required to store water for future use, to stop sea-water intrusion of a freshwater aquifer, to treat or renovate waste waters or sewage effluents, and to avoid or limit settlement of structures due to a lowering of the water table brought about by pumping from an adjacent construction dewatering scheme.

asphalt Asphalt comprises well-graded aggregate with a bituminous binder and filler, the latter consisting of inert material finer than BS 200 sieve size and usually comprising cement, limestone dust or fly-ash. Asphalt occurs naturally in asphalt lakes and in deposits mixed with sandstone and limestone rocks. Also obtained artificially in the refining of some hydrocarbons.

association, landform A grouping of geomorphological elements that is considered to make up a repetitive unit, from either an empirical or a genetic point of view. Used mainly to generalise areas of topography in order to produce maps for planning or engineering usage.

association, sediment A grouping of sediment types that is considered to make up a repetitive unit on either an empirical or a genetic basis. The concept is closely related to that of a facies association—i.e. an association of sediment types that is related to the juxtaposition of depositional environments in which the sediments accumulated.

ASTM American Society for Testing and Materials.

ATNAV Acronym for Acoustic Transponder NAVigation system. This survey-positioning system provides accurate offshore location control for surface vessels, submersibles, towed equipment and sea-bed structures. The system comprises an acoustic transducer aboard a vessel or platform which transmits signals and receives replies from a number of transponders on the sea-bed or elsewhere. The data are converted into position co-ordinates which can be presented in digital, analogue, paper print-out or cassette-tape storage format.

Atterberg limits and soil consistency In general, the physical properties of fine-grained cohesive soils are directly related to their moisture contents and four states of consistency can be recognised—i.e. liquid, plastic, semi-solid and solid. In 1911 a Swedish soil scientist, A. Atterberg, developed simple tests to determine the plasticity of soil for agricultural purposes, and these, known as Atterberg limit tests or *index properties,* have been widely adopted for engineering soil classification purposes. They include the liquid limit and the plastic limit, which define the water contents at the upper and lower boundaries of the plastic state as follows.

The *liquid limit (L L)* is the water content expressed as a percentage of the dry weight of soil when the soil passes from the liquid to the plastic state—i.e. when the soil first shows a small but definite shear strength. The test is carried out using either the *Casagrande liquid limit apparatus* or, preferably, the *cone penetrometer.*

The *plastic limit (PL)* is that water content expressed as a percentage of the dry weight of soil when the soil becomes too dry to be in a plastic condition and becomes semi-solid. This test is performed by hand-drying the soil and rolling it until threads 3 mm in diameter just become too dry to be in a plastic condition.

Shrinkage limit (SL) As the water content is reduced below the plastic limit, the soil first becomes a semi-solid and eventually reaches a solid state, when no further shrinkage will occur. The water content at this point is called the shrinkage limit and further reduction in water content is not accompanied by a decrease in the volume of the soil mass. Below the shrinkage limit, the soil is considered to be a solid. The test is now carried out by placing soil in a standard mould and measuring the linear shrinkage after drying.

The *plasticity index (PI)* is the numerical difference between the liquid and plastic limits, and represents the range of moisture content at which the soil is plastic. In combination with the liquid limit, the plasticity index gives an indication of the sensitivity of the soil to changes in moisture content.

The *relative consistency (C$_r$),* or *consistency index (CI),* defines the moisture content *(m)* of the soil in relation to the liquid limit and the plastic limit:

$$\frac{LL-m}{LL-PL}=\frac{LL-m}{PI}$$

For $C_r = 0$ the soil is at its liquid limit; for $C_r = 100$ per cent the soil is at its plastic limit.

The *liquidity index (L I)* is defined as the moisture content of the soil in excess of the plastic limit expressed as a percentage of the PI:

$$LI=\frac{m-PL}{LL-PL}=\frac{m-PL}{PI}$$

This merely describes the moisture content of the soil with respect to its index limits and is of no use for classification purposes. It is sometimes convenient to assess the condition of soil at its natural moisture content in the field.

In the Casagrande apparatus used for determining liquid limit, the clay is placed at four different moisture contents in a cup which is allowed to drop repeatedly onto a hard rubber seating until a standard groove, cut in the clay, just closes. The moisture content at 25 blows interpolated from a plot (flow curve) of blow count against moisture content gives the liquid limit. The *flow index* I_f expresses the relationship between change in moisture content and the corresponding change in shear strength, and equals the slope of the flow curve. The ratio between plasticity index and flow index is termed the *toughness index* I_t. These terms are now little used.

attrition   The resistance of one surface to the motion of another surface rubbing over it. *See Deval attrition test*.

attrition test   *See Deval attrition test*.

^{198}**Au**   *See gold-198 (^{198}Au)*.

auger, hand   *See hand auger*.

augered piles   *See Franki piling systems*.

auger, power   A machine-driven auger up to about 1.25 m diameter for drilling deep holes in the ground. Normally built onto a transporter vehicle.

aureole grouting   Aureole or umbrella grouting is a technique involving drilling and grouting ahead of a tunnel face to allow excavation to proceed in otherwise unstable ground.

autochthonous   In place, or not having underground transport. Applied originally in areas of tectonic deformation to beds that could still be traced to their supposed site of deposition (*allochthonous*), but now used also of sediments or weathering products that have accumulated at the place of production.

automatic engineer's level   A level which automatically establishes the horizontal line of site by means of a compensator suspended on the telescope. Precision to ± 1 second of arc is obtainable.

automatic settlement plotter   An instrument marketed by Soil Instruments Ltd, which allows of continuous measurement of elevation along a length of small-bore tubing, up to 1200 m long, laid to form a continuous horizontal loop in a foundation or structure during construction. A mercury–water surface is formed and pumped along the tube at a fixed rate; by monitoring the hydraulic head difference produced at the interface, a continuous record of interface elevation is obtained.

automatic tide gauges   Instruments to allow of automatic measurement of the level of the sea surface, including: *Float-operated gauge*, whereby a float is directly connected to a chart recorder by a wire.

Pressure-operated gauge, whereby the tide exerts a pressure on a submerged diaphragm which actuates the recording mechanism. *Bubbler gauge*, wherein a constant flow of compressed air is released via a bubble chamber to an open ended tube set at a known depth below the lowest astronomical tide. The pressure variation above the open tube due to the tidal variation is recorded on a drum chart. *Electrical capacitance gauge*, which uses the varying electrical capacitance of a wire mounted vertically in the intertidal zone. *Electrical resistance gauge*, similar to the capacitance gauge but using the changing resistance between two parallel electrical wires to determine water level. *Inverted echo sounder*, which is laid on the sea-bed and measures the reflection of the signal transmitted to the surface.

automatic trip-hammer or monkey   A heavy weight used for driving casing, etc., which can be lifted and automatically released at a preset height. An automatic trip-hammer having a mass of 63.50 kg (140 lb) is standard equipment for driving the *standard penetration test* sampler.

Autotape   A marine survey instrument for distance measurement similar to the *Hydrodist* but using a single interrogator unit instead of two master units in the survey vessel, and usually with an omnidirectional whip antenna which automatically interrogates one or two responders at co-ordinated shore stations. *See also Tellurometer*.

average stress, or octahedral normal stress   The average of the *major principal stress (σ_1), intermediate principal stress (σ_2) and minor principal stress (σ_3)* (i.e. $(\sigma_1 + \sigma_2 + \sigma_3)/3$). *See also intermediate principal strain; major principal strain; major principal strain; octahedral shear stress.*

average velocity   A hydrological term defining the volume of water that passes through a unit cross-sectional area of ground divided by its porosity.

axes of principal stress   *See principal stresses.*

axial plane   The plane which occurs through a fold across which the dip of the limbs is reversed. This may or may not be a plane of symmetry.

axis, fold   A line contained in the axial plane of a fold; hence, a line showing the direction of strike of a fold.

B

back-acter, or hoe   A type of excavator having a bucket on the end of a jointed, normally hydraulically operated arm which digs back and mainly below the machine. Employed for loading spoil into transport, digging trenches and trial pits—particularly for geotechnical site investigations.

back-prop A raking prop used to transfer the weight of timber to the ground in deep trenches. Usually placed below every second or third frame.

back-tilt The attitude assumed by material that has been involved in a circular landslip (*see rotational slip*). Often the soil or rock retains its cohesion, and rotates as a single unit. In these circumstances, the rear of the block sinks, so back-tilting the original ground surface.

bacteria In water wells, bacteria are micro-organisms that develop and multiply but do not cause disease. Common types include the genera *Clonothrix*, *Crenothrix* and *Gallionella* and the species *Leptothrix siderocapsa* and *L. sphaerotilus* (all aerobic), and sulphate-reducing bacteria (anaerobic).

bailer, or shell An open length of tubing fitted with a lifting bail at the top and either a flap, clack, dart or dump valve at its lower end. The tube is surged at the base of a borehole to collect sand, rock chippings, etc., which can then be lifted clear of the hole and disposed of.

baling-out permeability test . A simple field method of determining *permeability*, whereby water is baled out of a *borehole* to depress the water level and either readings are taken of the water level at various time intervals as it is allowed to rise and equilibrate (*rising-head permeability test*) or the water is maintained at a constant level and the rate of outflow measured (*constant-head permeability test*). The calculation of permeability *(k)* for these types of test is dependent on a knowledge of the borehole cross-section *(A)*; a shape factor *(F)* depending on the shape and dimensions of the intake and the soil profile; the *basic hydrostatic time lag (T)*, being the time required for equalisation of the pressure difference *(H)* existing at maximum drawdown between the original ground-water level and the water level in the borehole; and the quantity of outflow *(Q)*. Theoretical derivations of *k* for various shape factors using the relationship $k = A/FT$ or $Q = FkH$ are given by Hvorslev *(Figure B.1)*. *See field permeability tests*.

Balkan piling system This system incorporates square-section precast-concrete units in standard lengths of 5–13 m joined together by special steel joints. Nominal pile loads of up to 550 kN safe working load (235 mm side) and 500–1000 kN safe working load (275 mm side) are available, the latter size also being available with 42 mm inspection tubes passing down the full length of the pile through special joints to enable verticality and continuity to be confirmed by plumbing. Pile lengths up to 100 m are possible, depending on substrata conditions.

ball structures Structures that form at the interface of two dissimilar materials as a result of loading. They are from the intrusion of

20

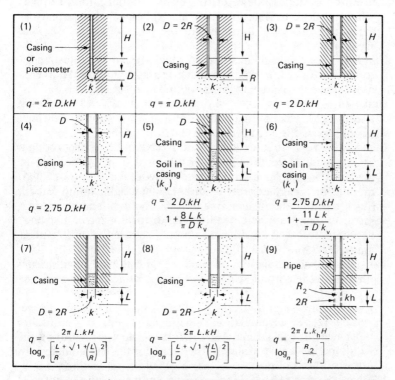

q = rate of flow in cm³/sec, H = head in cm, k = coefficient of permeability in cm/sec, $l_n = \log_e$, dimensions in cm, k_v = vertical permeability, R_2 = effective radius to source of supply well point

Cases (1) to (8) uniform permeability and infinite depth of pervious stratum assumed.

1 Spherical intake or well point in uniform soil
2 Semi-spherical soil bottom at impervious boundary. (Based on (1) or Dachler)
3 Soil flush with bottom at impervious boundary. (Formula by Dachler)
4 Soil flush with bottom in uniform soil. (Empirical data by Harza Taylor)
5 Soil in casing with bottom at impervious boundary. (Based on (3))
6 Soil in casing with bottom in uniform soil. (Based on (4))
7 Well point or hole extended at impervious boundary. (Approximate formula by Dachler cylinder replaced by semi-ellipsoid)
8 Well point or hole extended in uniform soil. (Approximately based on (7), cylinder replaced by an ellipsoid)
9 Well point through permeable layer between impervious strata

Figure B.1 Inflow and shape factors (Hvorslev, 1951)

21

mushroom-shaped bodies of the lower material into the upper: when seen in vertical section, the heads of the mushrooms appear as detached balls. Also known as ball and pillar structures. *See load structures.*

bankfull The condition of a river in which the banks are just about to be overtopped. It is believed that in this condition the river is most active in producing geomorphological changes.

bank storage The storage of water in permeable strata lining streams and rivers during periods of high-level water flow, which discharges as the stream/river level drops.

Banut piling rig An integrated hydraulic piling rig incorporating a hydraulically lifted drop-hammer mounted on a hydraulic base machine. Principal features include: hydraulic operation, high mobility and manoeuverability on site, light weight, low noise level, quick erection and dismantling on site (about 30 minutes by two men), automatically actuated hammer drop allowing of frequency variation from 50 to 80 blows/min. Produced by AB Anläggningsutrustringar AB of Sweden.

bar (1) An accumulation of sand or gravel that results from fluctuations in current velocity in a river channel. The growth and erosion of bars is considered by some to be the basic mechanism of river channel migration and, hence, of flood plain activity generally. (2) An accumulation of sand or gravel due to offshore movement of coastal material. Such bars often grow during the winter period, and lose material onshore during the summer. (3) In physics, a unit of pressure: 1 bar $= 10^5$ N/m$^2 = 14.5$ lbf/in^2.

barchan A type of sand dune, crescentic in plan, with the horns of the crescent facing downwind.

barium sulphate *See barytes.*

barometer An instrument used for measuring atmospheric pressure. In the mercury barometer the pressure of the atmosphere is balanced against the weight of a column of mercury, the two main types being the Fortin and the Kew. The aneroid barometer contains one or more thin metal capsules, partially evacuated of air. The change in the thickness of the capsules is magnified and typically is represented by a needle and dial. Precision aneroid barometers are sensitive to 0.1 mb, with an overall accuracy of ± 0.3 mb.

barrel Equal to 42 US gallons or 34.972 Imperial gallons.

barrettes Foundations constructed by trenching in weak ground by such techniques as those used for *diaphragm-walls.*

barytes Barium sulphate: the commercial product contains small amounts of iron oxide, silica and other minerals, and is used in drilling muds to make them heavier.

basalt A general name for a family of extrusive igneous rocks, characterised by the association of calcium-rich plagioclase feldspar with mafic minerals, particularly pyroxenes. Two principal varieties exist: olivine basalts, which are undersaturated with quartz and, hence, in which olivine is found; and tholeiitic basalts, which are quartz-saturated and, hence, in which the mafic minerals are dominantly pyroxenes. In general, olivine basalts are formed by the partial melting of the mantle material under high pressures, and so are characteristic of continental regions where the crust is thicker; tholeiitic basalts form by melting of the mantle under lower pressures, and so are characteristic of areas of this oceanic crust. The coarse-grained equivalent of basalt is termed *gabbro*.

basal till   *Till* that has been deposited from the base of a glacier or ice-sheet. Synonymous with lodgement till or ground moraine.

base exchange   Certain materials with large specific surfaces (surface area per unit volume) are able to adsorb ions onto their surfaces. For example, clay minerals are particularly noted for their ability to absorb cations. When in contact with a solution containing a particular cation, clays will adsorb that cation and release other cations that are already held. This phenomenon is termed *base (or cation) exchange*, and is reversible. A cation exchange capacity can be defined for clay minerals which reflect the amount of available surface and, hence, reflects the mineral structure. Kaolinites with a non-expanding structure have low exchange capacities, whereas expanding clays such as montmorillonite have high exchange capacities (about ten times that of kaolinites).

base level   The notional level to which river courses are supposed to adjust over a period of time. In many cases this will be sea level, but lake levels are also involved. The term is often used in the context of 'change of base level', when there has been a change in the activity of a river system as a result of a change in sea or lake level.

basement   A term used in regional geology for old crystalline rocks on which are located younger sedimentary rocks. Often also referred to as a 'basement complex', where the implication is that an earlier history of deposition and (usually) metamorphism has preceded the deposition of one or more series of sedimentary rocks.

basement complex   *See basement.*

BASIC   A computer-language acronym for Beginners All-purpose Symbolic Instruction Code. It is used mainly for *online* programming.

basic hydrostatic time lag   The time (T) required for equalisation of the pressure or head difference (H) between standing groundwater level and the water level in a well of cross-sectional area A, whereby $T = A/(Fk)$ at the initial rate of flow, $q = FkH$, where $F =$ the shape factor for the well, dependent on its geometry and boundary conditions controlling entry of water into the well, and k is the coefficient of permeability of the ground. The theory of basic time lag is discussed by M. J. Hvorslev in 'Time lag and soil permeability in groundwater observations', *Bulletin No. 36*, Waterways Experiment Station, Vicksburg, Mississippi, April 1951, and inflow and shape factors for a range of boundary conditions are given in *Figure B.1*.

basic rocks   A general term for those igneous rocks in which the ferromagnesian minerals are dominant. The term derives from the concept that there is a balance between metal oxides (bases) and silica (acid).

basin   (1) In sedimentology, an area which has acted as a locus for the deposition of sediment. At the time of deposition the area formed a local topographic depression. (2) In structural geology, a downwarp in a fold bed.

basket core lifter   A core-retaining device comprising a number of flexible spring metal fingers attached to a steel ring.

batch system   A method of grouting by injection whereby chemicals are mixed together in a single container before passing to the injection pump.

batholith   A large intrusive body of igneous rock, usually granitic. Batholiths are distinguished from other intrusions by their generally larger size and their equidimensional shape.

BAT piezometer   Manufactured by Linden-Alimak, Sweden, the BAT piezometer comprises a filter tip screwed into 25 mm galvanised pipe, which is driven down into the ground to a predetermined depth, and separate sensor equipment which can be lowered down the pipe after installation, the pore pressure being measured by a portable digital read-out instrument. Advantages include the ability for one measuring unit to handle all the piezometers installed and for the pipe to be capped at ground level to minimise damage.

BAT pore-pressure probe   A device for determining the geological profile of soft soils by recording the pore pressures generated by a probe as it penetrates the soil at constant speed. Manufactured by Linden-Alimak, Sweden, the equipment is portable and comprises a pushing device, a power pack, a pore-pressure sounding tip with connecting pipes and a self-contained automatic read-out unit. The probe can identify thin seams of sand, silt or clay embedded in a soft soil deposit by measuring changes in pore pressure.

bathythermograph   An instrument used for marine conditions to determine temperature—depth profiles.

batter   An artificial slope, either excavated or constructed, which has a uniform angle. Also used for the angle itself; thus, 'a batter of 1 (vertical) in 2 (horizontal)'. *See slope angle.*

batter pile   *See raking pile.*

Baume gravity   A measurement of the *specific gravity* of a liquid, obtained by use of a graduated float hydrometer:

$$\text{specific gravity} = \frac{145}{145 - \text{Baume gravity}}$$

Heavy Baume is for liquids heavier than water. The specific gravity of water at 4° C is equivalent to zero degrees Baume. 100 degrees Baume represents a specific gravity of 3.222.

beach profile The cross-sectional shape of the beach seen in vertical elevation. Typically there are seasonal changes in profile due to differences in wave activity: constructive periods of (quiet) activity lead to a build-up of material, whereas destructive periods of (storm) activity cause the loss of material, except that thrown up to form a storm beach. The material lost during the latter periods often accumulates as offshore *bars*.

beam-type gravimeter *See gravimeters.*

bearing capacity Bearing capacity is the ability of the ground to accept loading. Thus, the ultimate bearing capacity is the loading intensity on a given size of foundation which just causes shear failure of the ground, and the *safe bearing capacity* is the loading intensity which allows of a suitable factor of safety against shear failure. The latter is not the same as **allowable bearing pressure**. Either the *ultimate bearing capacity* or the *safe bearing capacity* can be quoted as gross or net values which either do or do not take account of overburden pressure. These can be simply illustrated by the following equations for a purely cohesive soil ($\phi = 0$):

ultimate bearing capacity $(q_f) = C . N_c + p$
net ultimate bearing capacity $(q_{nf}) = C . N_c$
safe bearing capacity $(q_s) = \dfrac{1}{F} C . N_c + p$

net safe bearing capacity $(q_{ns}) = \dfrac{1}{F} C . N_c$

where C = cohesion or shear strength; N_c = bearing capacity factor; p = overburden pressure; and F = factor of safety.

bearing pile A pile which transmits the weight of a structure to the ground; termed an end-bearing pile if the load is supported mainly by resistance developed at the pile base, or a friction pile if supported mainly by skin friction developed along the pile shaft.

bed The basic unit of sedimentary rock. A bed is a lithological entity that is formed as a result of a single depositional process or episode. In the field, a bed is separated from its neighbours by contacts or bedding planes across which there is a structural or lithological break.

bedding plane   The boundary surface of a bed, which results from a local period of erosion or non-deposition.

bedding plane slip   Movement along a bedding plane which allows relative movement between beds to be accommodated. Such movement arises either from structural disturbance or from compression within the beds during lithification.

bed load   The material transported in a moving fluid by the process of traction, as opposed to suspension. The bed load thus consists of the larger particles carried by the flow, which move by sliding, rolling and bouncing along the bed under the influence of the shear stress exerted by the moving fluid.

beheading   The capture of the headwaters of one river by another, so shortening its length and causing the headwaters to become tributary to the capturing river.

bell pit   A characteristic excavation made for winning coal at its outcrop during late mediaeval and Tudor times, comprising a shallow shaft (seldom more than 10 m deep) expanded at its base by undercutting in the seam to give a bell-shaped cross-section. Short radial headings were often driven from the base of the pits, which were usually very closely spaced along the down-dip edge of the outcrop.

bench-mark   A fixed point of reference with a known elevation, established in a mapped country at intervals throughout the country by the State, which publishes the position and elevation above the standard datum. In Britain *ordnance bench marks* (OBM) are chiselled on buildings and milestones in the form of an arrow surmounted by a horizontal groove, the centre-line of which defines the elevation.

Benoto piling system   Used in the UK since 1955, this system is operated by Lilley-Waddington Ltd and allows of construction of cast-in-place reinforced concrete piles with diameters up to 1180 mm to a maximum depth of 45 m carrying loads up to 5500 kN/pile. Piles can be raked up to 15° to the vertical. The system provides for a rigid temporary casing to the full depth of the pile, which is hydraulically operated and jacked down while excavation proceeds within the casing by hammer-grab. The excavation method allows the pile to be installed in difficult ground conditions, including boulder clays, rocks and granular fill. The casing is withdrawn hydraulically as the concrete is placed, compaction of which is assisted by oscillation of the casing. The equipment is also used for the construction of *secant (interlocking) pile* walls for subsurface structures such as basements, cofferdams and retaining walls. These are formed by constructing two female piles and then a subsequent interlocking male pile by cutting back the concrete of the females. All the piles can be reinforced as required.

26

bentonite A clay derived from weathering of volcanic dust and ash deposits consisting mainly of montmorillonite characterised by a high *liquid limit* (350–500 per cent).

berm (1) A level platform cut or formed into the side of a cutting or embankment to increase the overall stability of the slope or to trap falls of rock or other debris from a high steeply cut face. Also formed in the slope at the junction of an impervious soil or rock with an overlying pervious material to allow of provision for a collecting channel or pipe drain for seepage flows. (2) An accumulation of material (for example, on a beach) that leads to a break in the slope profile.

Bernoulli's theorem *See total energy of a fluid.*

beta (β) method of pile design An effective stress method used to determine the average unit skin friction between mainly cohesive soil and pile shaft in a soil layer of thickness i, whereby

$$f_{si} = k \tan \delta p_0' = \beta p_0'$$

in which k = the average coefficient of earth pressure on the pile shaft; $\tan \delta$ = the average coefficient of friction between the soil and the pile shaft; and p_0' = the effective overburden pressure. In normally consolidated clays, when no appreciable change in lateral ground stress conditions occurs, $k = 1 - \sin \phi'$ (where ϕ' is the angle of internal friction in terms of effective stress) may be assumed. *See also alpha (α) method of pile design; lambda (λ) method of pile design.*

Reference: 'Bearing capacity and settlement of pile foundations', G. G. Meyerhof, *Proc. Am. Soc. Civ. Eng.*, **102**, No. GT3, March 1976.

'B' horizon The lowest part of the soil profile, where part of the substances washed out from the overlying 'A' horizon are precipitated out and accumulate. *See 'A' horizon; 'C' horizon.*

biat (byatt) A wooden bearer used to support guard rails, walkways and decking, etc.

bid The offer or proposal submitted by the *bidder* giving the prices for the work to be executed.

bidder A person or organisation submitting a *bid* for execution of a contract.

bing A Scottish term for colliery spoil heap resulting from the winning of coal by mining.

Bingham substance An idealised model material which has both a finite yield stress (S_0) and a strength component that is proportional to the rate of strain ($\dot{\varepsilon}$). The constant of proportionability (η) is known as the *Bingham viscosity* of the material. The stress–strain rate behaviour of such a substance is given by the equation

$$T = S_0 + \eta \dot{\varepsilon}$$

where T is the shear strength of the material.

Bingham viscosity *See Bingham substance.*

birdsfoot delta The name given to a type of delta in which the distributary channels, flanked by levées, extend beyond the main delta, so producing finger-like outgrowths resembling a bird's foot. The Mississippi delta is a classic example of this type.

Bishop consolidometer A *soil mechanics* testing machine designed by Professor A. W. Bishop of Imperial College, London. A thin circular sample of soil is contained within a metal ring and is subjected to a uniaxial compressive stress via a system of dead-weights and levers. The time-dependent compression of the material is recorded in order to determine its *consolidation* and *compression* characteristics. The machine is also commonly termed an *oedometer.*

Bishop method of slope stability analysis This method is similar to the *Swedish circle method of slope stability* in that it divides the potential failure mass into a number of vertical slices *(Figure B.2).*

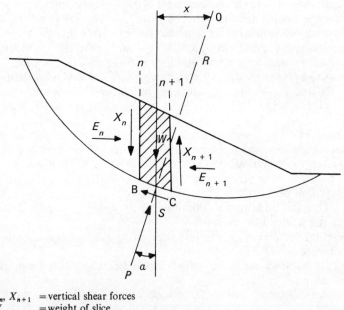

X_n, X_{n+1} = vertical shear forces
W = weight of slice
P = total normal force acting on base
S = shear force acting on base
α = angle between base and horizontal
E_n, E_{n+1} = resultants of horizontal forces acting on sections n and $n+1$

Figure B.2 Bishop's method of slope stability

28

However, slopes containing several different soil types can be accommodated, as can the pore-pressure regime, to allow of an *effective stress* analysis for long-term stability. Oblique side forces on each slice are considered, as against the assumption that the lateral forces are equal on both sides of each slice in the Swedish circle method. The forces acting on each slice are determined from the limit equilibrium of the slices and the equilibrium of the whole obtained by summation of the forces on all the slices.

Reference: 'The use of the slip circle in the stability analysis of slopes', A. W. Bishop, *Géotechnique*, V, No. 1, pp. 7–17, 1955.

Bishop and Morgenstern method of slope stability analysis   Based on the *Bishop method of slope stability*, this method considers the pore-water pressure ratio, r_u, and gives the factor of safety (F) of the slope as

$$F = m - (n)(r_u)$$

where m and n are stability coefficients. r_u is defined as $r_u = \mu/\gamma h$, where μ = pore water pressure; γ = soil density; and h = depth of point in soil mass below ground level. The stability coefficients may be obtained from a series of charts giving various values of:

(a) slope angle β
(b) $c'/\gamma H$ where c' = effective cohesion; γ = soil density; and H = height of slope
(c) ϕ' = effective angle of shearing resistance
(d) D, defined by Taylor as the depth to a hard stratum divided by the height of the slope.

Reference: 'Stability coefficients for earth slopes', A. W. Bishop and N. R. Morgenstern, *Géotechnique*, X, No. 4, pp. 129–150, 1960.

Bishop sampler   A soil sampler that incorporates the use of compressed air to retain soil samples such as sand and very soft clay that would otherwise tend to drop out of the sample tube—especially when taken below the water table. Samples are normally of 60 mm diameter and can be recovered in a reasonably undisturbed state provided that a static thrust is used to push the sampler into the ground.

bit   (1) A binary digit, and the smallest unit of information that can be fed into a computer to allow a choice to be made between, say, one/zero, on/off, yes/no, stop/go. (2) The tool on the bottom of a drill string that cuts into the soil or rock during drilling—e.g., diamond bit, rock-roller bit.

bitch   A metal fastening used to secure two timbers that cross each other. Similar to a *dog* but having only one end at right angles to the other.

bitumen   A mixture of hydrocarbons of natural or pyrogenous

29

origin, which are soluble in carbon disulphide and which exist in the liquid, semi-solid or solid state.

bitumen binder   Bitumen binders can be of two types: (a) 'Straight-run', obtained by straight distillation of petroleum and stopping at a certain temperature. (b) 'Cut-back bitumen', obtained by adding a solvent to a straight-run bitumen. Different solvents, such as naphtha, kerosene and diesel oil, produce different curing rates.

bitumen emulsion   An emulsion of bitumen in water made by adding emulsion (e.g. resinous soap or caustic soda) to water and adding bitumen in an emulsifier mill. Cationic emulsion is made in the opposite way to the above method, resulting in bitumen containing globules of water.

bit wear tests   Tests devised to determine the abrasive resistance of rock by measuring the bit wear of a standard bit, drilling for a specified time, under specified conditions.

Bjerrum's danger limits for distortion of structures   The degree of angular distortion that a structure can undergo before causing distress to the building itself or that which would impair its functional ability (*see Table 13*).

Reference: 'Discussion on compressibility of soils', L. Bjerrum, *Proc. Europ. Conf. on Soil Mech. and Found. Eng.*, **2**, pp. 16–17, Wiesbaden, 1963. *See also Values of Limiting Differential Settlement*, USSR Building Code, 1955.

black cotton soil   A brown or black tropical heavy clay soil (40–50 per cent clay fraction) derived from basalt rock, which is characterised by large volume changes on swelling and shrinkage due to seasonal moisture content changes.

blading back   Respreading soil by grader over the area from which it was previously taken or set aside in windrows.

blaise   See *colliery shale*.

blanket course   A layer of free-draining material placed at the base of an embankment to permit of drainage and/or dissipation of excess pore-water pressures in the embankment and the underlying ground. Often used in conjunction with vertical *sand drains* installed below the blanket. Also used as a construction expedient to allow construction plant to work on soft ground, when it may also incorporate the use of a proprietary form of filter fabric.

blanket grouting   Blanket grouting involves drilling grout holes on a closely spaced grid pattern so as to produce a more or less even spread of grouted ground over a particular area or region.

blasting *(well stimulation)*   See *shooting or blasting*.

bleeder well   A well sunk to relieve pressure from an *aquifer* under *artesian* or sub-artesian pressure.

bleeding   A term describing the separation of water from a grout with the passage of time.

blinding concrete A thin layer of low-grade concrete placed on mainly cohesive soils as soon as possible after excavation reaches formation level, to protect them from surface water or rain-water and to provide a working surface for men and plant.

block disruption The break-up of landslide blocks by minor processes of erosion or movement, leading to the production of *colluvium*.

block faulting A pattern of faulting in which a series of displaced units or blocks is created.

block field A relatively horizontal area that is strewn with large rock fragments derived by the break-up of the local bedrock due to frost action. Block fields are characteristic of mountainous or high-latitude areas where freeze–thaw processes are active.

blocking, chock or chog A timber block used as distance piece or packing (e.g. between a waling and the temporary or permanent lining of an excavation) to permit of the insertion and erection of vertical reinforcement in retaining walls or other permanent constructions.

Reference: BS 6031 : 1981.

block sample A sample cut by hand from material exposed in an excavation, and normally taken in rock and cohesive soil. Often taken to provide a specimen containing part of a slip plane.

block size (resulting from discontinuities) One of the ten parameters selected to describe discontinuities in rock masses, being the rock block dimensions resulting from the mutual orientation of intersecting joint sets and resulting from the spacing of the individual sets. Individual discontinuities may further influence the block size and shape. *See discontinuity*.

Reference: International Society for Rock Mechanics, Commission on Standardisation of Laboratory and Field Tests, 1977.

blooey-line A duct or pipe to take cuttings away from the top of a borehole when drilling with air or gas.

blow, or boil Displacement of soil in the base of an excavation caused by upward flow of water due to differential head.

blown sand Sand deposited after transport by wind—e.g. dune sand. Grains can become nearly spherical in shape.

blow-out Uncontrolled sudden flow from an oil or gas well.

boart-bortz A fairly low-quality industrial diamond.

body waves Those elastic waves that travel through the body of the transmitting medium, and thus are unaffected by the presence of the boundaries.

bog Very soft waterlogged ground, often containing highly organic soils such as peat.

boiling *See critical hydraulic gradient*.

31

bonding The adhesion of cement or glue to a surface. Chemical bonding is the capacity of atoms to combine and become molecules.

bonds Bid, performance and payment bonds and other instruments of security, furnished by the contractor and his surety in accordance with the contract documents.

boomers Instruments used in marine geophysical surveying to provide a seismic-energy source by charging capacitors to high voltage and then discharging them through a transducer in the water. *See continuous seismic reflection profiling.*

bord and pillar *See subsidence.*

bored and cast-*in-situ* concrete piles These piles are formed by excavating or boring a hole in the ground, with or without *casing*, and filling with plain or reinforced concrete, or with precast concrete sections which are grouted in place. Such piles were introduced in the UK in 1919, when F. Smith & Son (Grimsby) Ltd constructed 50 No. 305 mm diameter piles through 9.14 m of soft *alluvium* into *boulder clay* and filled them with concrete after inserting reinforcement cages. These were estimated to carry 200 kN each. This company eventually became the Expanded Piling Co Ltd, of Grimsby, who with other piling companies in this field can provide bored piles of up to 1800 mm diameter with under-reamed bases of 3 × shaft diameter to carry safe working loads of up to 15 MN. In addition to normal foundation piling, bored piles (generally 600 mm diameter size) can be used to form temporary or permanent basement walls by continuous construction around the required perimeter, and large-diameter piles can be used to form retaining walls. Also, large-diameter caisson piles of 600–3600 mm formed by rotary and vibratory methods with under-reamed bases are possible. These methods have been employed for shaft-sinking purposes connected with mining and drainage. Individual bored piles of 1800 mm diameter and belled out to 5400 mm diameter have been used to carry loads of up to 30 MN. *See also secant (interlocking) piles; short bored pile; pile foundations.*

bored piles *See bored and cast-in-situ concrete piles.*

borehole A hole formed in the ground by hand- or machine-boring methods. Site investigation boreholes are generally undertaken in superficial deposits by *shell and auger boring* methods, which allow undisturbed samples to be taken for laboratory analysis or *in situ* testing of the materials in the borehole. In solid strata *rotary drilling* methods allow core samples to be obtained for testing and geological inspection.

borehole caliper A device comprising three spring-loaded arms set at 120° intervals for lowering down a borehole to obtain information on the variation in its diameter with depth. Experienced geophysicists can interpret the information to aid stratigraphic correlation and to identify lithology. It can also be used to determine under-reaming, to detect casing separation and to measure the effects of dynamiting.

borehole gravimeter A remote-reading *gravimeter* that can be lowered down a borehole to obtain gravity readings at any specific depth.

borehole impression device A device marketed by Triefus Industries Ltd that allows a thermoplastic film mounted on thin metal plates to be pushed against the sides of a borehole by inflating a rubber tube surrounding a perforated metal tube. An impression of any *joint*, *discontinuity*, etc., is obtained and an instrument can be incorporated to record the orientation of the impression.

borehole jack A device for performing small-scale load tests in boreholes developed by Wimpey Laboratories Ltd, England, to obtain information on the strength and deformation characteristics of rocks and hard soils. The instrument is designed to fit closely in a 146 mm (SF) diameter borehole incorporating a transverse jack which can impose pressures of up to 25 MN/m^2 on the borehole wall, the pressures and ram travel being controlled and monitored by a console operated at ground level.

borehole pressure recovery test A *permeability* test for execution in boreholes up to about 230 m depth in material having a permeability of around 10^{-5} to 10^{-7} m/s, whereby water is pumped from a sealed-off test section and the discharges measured for various pressure reductions. At maximum steady state drawdown, pumping is discontinued and a plot of the pressure recovery with time obtained from which the formation permeability can be calculated.

borehole record A record of the soil–rock profile at the particular location of a borehole. It should be in accordance with a recognised classification system (e.g. BS 5930: 1981, Code of Practice for Site Investigations—formerly British Standard Code of Practice 2001, Site Investigations) and be based on the field descriptions given by the driller, amplified and modified by a geotechnical engineering description of any samples and rock cores taken, the results of *in situ* and laboratory tests and any known geology, for the site. The record should also include the ordnance datum level of the borehole location, depths and types of samples, details of *in situ* testing, type of boring and equipment used, details of casing required, dates of field work, borehole number and name of site. Typical borehole and trial pit records are given in *Figures B.3* and *B.4*, respectively.

borehole scanner, or televiewer An instrument that emits a narrow

Boring method	SHELL AND AUGER TO 5·80 m / ROTARY CORING TO 9·50 m						Location		Record of BOREHOLE 235	
Boring diameter (mm)	200 TO 2·55m; ·50 TO 4·30m / H TO 9·50m								(sheet / of /)	
Casing diameter (mm)	200 TO 2·55m; 150 TO 3·00m / H TO 6·00m						Orientation		Ground level (m.O.D.) 113·85	
Boring equipment	PILCON 20 BOYLES BBS 37 / HWF BARREL DIAMOND SET BIT								Date commenced 10/6/73	

Samples and in situ tests		Casing Depth (m)	Water Depth (m)	TCR SCR	RQD	Date and Depth (m)	DESCRIPTION OF STRATA	O.D. Level (m.O.D.)	Legend
Depth (m)	Type								
						10/6	FILL (CONCRETE, ASHES AND BRICKS)		
0·75	S33	0·60				0·60		113·25	
1·40	C26	1·40					MEDIUM DENSE SAND AND GRAVEL BECOMING CLAYEY BELOW 2·50m		
1·55	BD								
1·90	W		1·90*						
2·55	U↑	2·55							
2·55	BD					2·55		111·30	
			1·55			7·80			
			1·70			11/6			
3·00	U	3·00					STIFF TO VERY STIFF RED-BROWN SILTY CLAY, SANDY ZONES BELOW 4·10 m		
3·45	D								
4·00	U*	3·00							
4·30	D					4·30		109.55	
			1·70			12/6			
4·95 – 5·05	C50◉	3·00					WEATHERED BROWN FINE WEAK TO MODERATELY STRONG MICACEOUS SANDSTONE, JOINTED		
5·80	C50✱	3·00				5·80		108·05	
			1·70			17/6			
		6·00							
				80					
7·00	I₃ 20				20		SLIGHTLY WEATHERED MEDIUM BEDDED LIGHT BROWN FINE MODERATELY STRONG MICACEOUS SANDSTONE; STEEPLY DIPPING FISSURE WITH IRONSTAINED SURFACES 7·50m to 8·50m		
8·00	I₅ 15			43					
						8·80			
				90	45				
				60		9·50		104·35	
							END OF BOREHOLE		

REMARKS
HAND EXCAVATED TRIAL PIT TO 0·60m (16h)
BOREHOLE WAS ADVANCED BY CHISELLING FROM 4·30m to 5·80m (6h)
ON COMPLETION THE BOREHOLE WAS BACKFILLED WITH GRAVEL UP TO 1·00m
WITH A STANDPIPE INSTALLED AT 3·60m AND THE BOREHOLE SEALED AT
GROUND LEVEL. A STEEL PROTECTIVE COVER WAS CONCRETED INTO PLACE
OVER THE BOREHOLE.

For explanation of symbols and abbreviations see Notes, pages (i) and (ii)

LAB Ref. No. S/20565	DENHAM AIRPORT	Fig. 35

Figure B.3 Typical engineer's borehole record (*Wimpey Laboratories Ltd*)

Method of excavation *MECHANICAL EXCAVATOR*			Ground level (m.O.D.) *114·85*					
			Location *TQ 0388*		Record of			
Dimensions of trial pit (m) *3 × 1*			Date commenced *10/6/73*		TRIAL PIT *1*			
Samples and in situ tests		Water depth (m)	Date and Depth (m)	DESCRIPTION OF STRATA			O.D. level (m.O.D)	Leg-end
Depth (m)	Type							
		−	10/6	FILL (ASHES, BRICK FRAGMENTS, SAND, AND SILTY CLAY)				
0·50	BD							
			0·80				114.05	
1·00	BD			COMPACT CLAYEY FINE TO MEDIUM SAND WITH OCCASIONAL FINE TO MEDIUM ANGULAR FLINT GRAVEL				
1·20	W	1·20*	1·20				113.65	
1·40	D							
2·00	BD			FIRM FISSURED BROWN SILTY CLAY WITH TRACES OF GYPSUM				
2·40	U		2·40				112.45	
2·85	D	2·90	3·00	STIFF FISSURED GREY SILTY CLAY			111·85	
				END OF TRIAL PIT				

Remarks:
   THE SIDES OF THE PIT REMAINED STABLE DURING EXCAVATION

For explanation of symbols and abbreviations see Notes, pages (i) and (ii)

Method of excavation			Ground level (m.O.D.)					
			Location		Record of			
Dimensions of trial pit (m)			Date commenced		TRIAL PIT			
Samples and in situ tests		Water depth (m)	Date and Depth (m)	DESCRIPTION OF STRATA			O.D. level (m.O.D)	Leg-end
Depth (m)	Type							

Remarks:

For explanation of symbols and abbreviations see Notes, pages (i) and (ii)

Lab. Ref. No. S/ 20565	*DENHAM AIRPORT*	Fig. 36

Figure B.4 Typical engineer's trial pit record (*Wimpey Laboratories Ltd*)

35

acoustic beam which scans the sides of a *borehole* in a tight helix as the instrument is raised or lowered. The reflected wave is displayed on a cathode ray tube which can be photographed to give a picture or log of the borehole showing cavities, fractures, etc.

borehole TV A closed-circuit television system that allows a small TV camera to be lowered down a *borehole* and the sides of the borehole to be observed on a monitor set at the surface. Suitable only in a dry uncased borehole.

borescope An optical instrument for inspecting and measuring internal structure and cracks, etc., in rocks as they appear in the walls of a *borehole*. Initially used for inspecting mining and tunnelling works, but also useful for geotechnical investigations and inspecting pipes or other openings.

boron A non-metallic solid obtainable in powder (amorphous boron) and crystal (adamantine boron) forms. Deleterious to some plants when in excess of about 2 mg/litre in irrigation water.

Borros point A rod which is fixed into the ground by flexible metal anchors for use as a fixed reference point for *settlement* measurements. The rod is driven into the ground, or installed in a prebored *borehole*, and is isolated by casing from the surrounding soil. *See Geonor settlement probe.*

borrow pit An excavation made to provide material for aggregate, filling or *rip-rap (protection for slopes)*.

bottleneck slides *See rotational slip.*

bottom heave The heave or lifting of the bottom of an excavation. This is due to the reduced vertical pressure in the soil below the excavation bottom resulting in (a) elastic expansion of the soil as the overburden soil is removed, (b) the reduced pressure allowing the soil to absorb water and swell and (c) plastic flow of the soil into the excavation if the *critical depth of excavation* is exceeded.

bottomset beds Horizontal or subhorizontal sedimentary beds that form in advance of a prograding delta. They are usually fine-grained and laminated. In the classic sedimentary sequence that is formed by an advancing delta, the bottomset beds will be subsequently overlaid by the inclined foreset beds of the delta front.

boudinage A type of structure formed when dissimilar beds are subject to unequal compressive stresses and undergo plastic yielding. The stronger bed(s) extend into a series of bulges (boudins) that are separated by thin rocks, while the weaker bed(s) infill around these bulges. The phenomenon is similar to the 'necking' observed when a ductile material is subjected to a laboratory tension test. Geologically, the structure implies that the conditions during deformation were such that the rock could exhibit ductile behaviour.

Bouguer anomaly The deviation from the expected gravity field that

36

remains after corrections have been made for the elevation of the recording station and the thickness and density of rock between it and the datum level. The Bouguer anomaly reflects contrasts of density in the rock below the station, and is compared with the computed effects of possible geometric structures and density contrasts that might be expected to occur, in order to establish alternative models of subsurface structure.

boulder   A rounded to sub-angular rock fragment greater than 200 mm in size. *See particle size.*

boulder clay   An English term often used to describe mainly unstratified glacial deposits, even though they do not always contain boulders, sometimes contain little or no clay and occasionally consist of *silt, sand* and *gravel*. The Scottish term *'till'* is preferred. *See also glacial deposits; drift.*

Bouma sequence   A term given to a sedimentary sequence that arises during deposition from a turbidity flow. When fully developed, the sequence contains a lowermost sandy unit with larger clasts, followed by sandy units that exhibit sedimentary structures indicative of decreasing water velocity. The sequence is completed by a fine-grained unit. The units are interpreted to represent the transition from the stony base of the flow through a decreasing flow regime in which sand is deposited, to the final deposition of the fine material. Bouma sequences have been recognised widely in ancient sediments, particular in flysch deposits, where repetition of the sequence is commonplace, suggesting an origin for the deposit as a series of turbidity flows.

Boussinesq theory   A theory expressing the distribution of stresses produced by a vertical point load on the surface of a semi-infinite, homogeneous, isotropic elastic medium. The derived equations relating the stress components shown in *Figure B.5* are as follows:

$$\sigma_z = \frac{Q}{2\pi}\frac{3z^3}{(r^2+z^2)^{5/2}} = \frac{Q}{2\pi z^2}(3\cos^5\theta)$$

$$\sigma_r = \frac{Q}{2\pi}\left(\frac{3r^2z}{(r^2+z^2)^{5/2}} - \frac{1-2\mu}{r^2+z^2+z\sqrt{r^2+z^2}}\right)$$

$$= \frac{Q}{2\pi z^2}\left(3\sin^2\theta\cos^3\theta - \frac{(1-2\mu)\cos^2\theta}{1+\cos\theta}\right)$$

$$\sigma_t = -\frac{Q}{2\pi}(1-2\mu)\left(\frac{z}{(r^2+z^2)^{3/2}} - \frac{1}{r^2+z^2+z\sqrt{r^2+z^2}}\right)$$

37

$$= -\frac{Q}{2\pi z^2}(1-2\mu)\left(\cos^3\theta - \frac{\cos^2\theta}{1+\cos\theta}\right)$$

$$\tau_{rz} = \frac{Q}{2\pi}\frac{3rz^2}{(r^2+z^2)^{5/2}} = \frac{Q}{2\pi z^2}(3\sin\theta\cos^4\theta)$$

where $\mu = $ **Poisson's ratio**. An alternative form of the first equation given above is

$$\sigma_z = \frac{Q\dfrac{3}{2\pi}}{Z^2\left[1+\left(\dfrac{r}{z}\right)^2\right]^{5/2}}$$

an expression which is used in the computation of vertical stress used in the analysis of foundation settlement.

Reference: *Application des potentials à l'étude de léquilibre et du mouvement des solids élastiques*, V. J. Boussinesq, Gauthier-Villars, 1885.

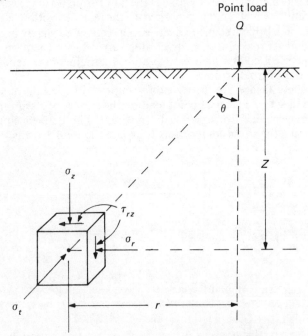

Figure B.5 Stresses in cylindrical co-ordinates caused by perpendicular point load at surface referred to in **Boussinesq theory**

Bowen reaction sequence Studies on igneous rocks indicate that the ferromagnesian minerals undergo a well-defined series of reactions as the temperature of the magma falls. At each stage silica is added from the magma to form a more complex chemical structure. The sequence is termed the *(discontinuous) reaction sequence*. The mineral families involves are olivine→pyroxene→amphibole→mica.

Bowen's reaction series Two series of minerals termed *continuous* and *discontinuous* by N. L. Brown, an American petrologist, giving the normal order of crystallisation from a melt.

box caisson *See caisson.*

box fold A fold structure in which an extensive horizontal crest is flanked by steeply dipping limbs.

box heading A small tunnel used in construction work, in which the roof and walls are close-timbered.

box pile A hollow pile formed by welding together rolled-steel sections. *See steel piles.*

box sextant A simple compact form of *sextant* used mainly in topographical survey work.

^{82}Br *See bromide-82 (^{82}Br).*

bracing The internal system of *struts* and *walings*, etc., used in *cofferdams, caissons* and other excavations to enable them to resist external pressures.

bradenhead A seal between two strings of casing in a borehole.

braided stream (river) A river system in which individual channels branch and rejoin, owing to the formation of *bars*. *See anastomosing.*

brake horsepower *See horsepower.*

Brazilian tensile strength test An indirect method of measuring *tensile strength* used in *rock mechanics*, whereby a disc of rock is placed between the platens of a *compression* test machine and loaded to failure across a diameter. To overcome the difficulty of assuring that the disc fails in tension, a central hole may be drilled to act as a stress concentrator.

breccia (1) A sediment consisting of angular fragments of broken rock. (2) (Fault) breccia: a zone of broken rock adjacent to a fault.

bridge An obstruction in a borehole due to caving of the borehole sides.

bridging time The length of time for which a freshly excavated span at the head of a driven tunnel can be relied upon to stand without support. Bridging times range from a few minutes for soft soils to years or decades for hard rocks.

British Soil Classification System (BSCS) *See soil description and soil classification.*

British Steel Piling Co Ltd cased pile This system uses a permanent steel *casing* formed from steel plate twisted into a continuous helix and welded. The casing has a flat plate welded to its base and is driven by an internal drop-hammer on to a plug of concrete cast at the bottom of the casing. The casing is hearted with concrete when the required depth is reached. Reinforcement can be added but is not normally required. Diameters from 250 mm to 700 mm are available to carry working loads from 150 kN to about 2500 kN.

The **BSP vibro pile** is an *in situ* concrete pile which uses a steel casing thickened at its lower end to form a tamping ring and fitted with a detachable cast-iron shoe. After driving to the required penetration and/or *set*, the casing is gradually filled with concrete while withdrawal of the casing and tamping of the concrete proceeds simultaneously, the withdrawal being effected by an alternating up and down movement of the pile-hammer attached to the casing through extractor links. The resultant shaft is corrugated to give high frictional resistance between it and the surrounding ground. The concrete can be reinforced if required.

A modification of the above is the '*double driving*' *method*. In this the pile is formed in the normal way (the concrete being unreinforced) and a further tube driven before the concrete has a chance to set. Concreting and reinforcing is then carried out as the casing is withdrawn. The method obtains a pile with up to 40 per cent more surface area.

See pile foundations.

brittleness index (I_B) This is the fall-off in strength from peak to the residual value. Thus

$$I_B = \frac{S_f - S_r}{S_f}$$

where S_f = peak shear strength of soil and S_r = residual strength of soil. *See residual and peak shear strength.*

brittle support   A brittle support is one which on failure may cause the immediate collapse of the support system.

brob, or nail spike   A metal fastener used in timbering, which has its head bent at right angles to the shaft.

Bromhead ring shear   Apparatus used in *soil mechanics* to determine residual strength of soils whereby a thin remoulded specimen is subjected to rotational shear. Drainage paths are very short, allowing of immediate dissipation of pore-water pressures.

bromine-82 (^{82}Br)   A radioisotope of bromine with a half-life of 35.7 h, used for radio-tracing.

Brownian movement   The erratic movement of small particles in a suspension caused by their bombardment by moving molecules of the solvent. Established by Robert Brown in 1827.

BS British Standard.

BSI British Standards Institution.

BSP vibro piles *See British Steel Piling Co Ltd cased pile; pile foundations.*

Btu British Thermal Unit: the heat required to raise the temperature of 1 lb of water through 1 °F.

bubbler tide gauge *See automatic tide gauges.*

bucket-wheel excavator A particular type of front-loading machine designed for continuous digging performance, comprising a revolving head with a number of buckets which dig at the end of an arm pointing away from the machine and load onto a conveyor belt stretching down the arm across the top of the machine and onto a secondary conveyor belt.

building owner A person or persons intending to demolish, erect, extend or alter a building or other structure. *See also adjoining owner.*
Reference: CP 2004 : 1972.

bulk density The mass of a material, including solid particles and any contained water, per unit volume, including voids.

bulk modulus *(K)* *See stress–strain relationships.*

bulldozer A wheeled tractor or crawler tractor provided at its front with a dozer blade set at right angles to the direction of travel.

bungum A term used in London for recent alluvial silt and generally applied to types of alluvial soft clay with or without silt.

buoyancy *See floating foundation.*

buoyancy raft A hollow sub-structure/box foundation designed to spread the building load onto the soil and of such a depth that the weight of soil removed for its construction is about the same as the combined weight of the superstructure and its supporting raft foundation. In this way the net load on the soil is reduced to a minimum.

buoyant foundation *See floating foundation.*

Burbank test A test designed to determine the relative abrasiveness of a rock sample on metal parts of mining and crushing equipment, wherein a metal paddle of the test alloy is counter-rotated inside a drum containing pieces of the rock to produce high-speed impact and rapid wear of the test paddle. *See also abrasion; abrasion test; Los Angeles abrasion test; sand blast test.*

Bureau of Public Roads (BPR) Classification System A system of soil classification developed by the US Bureau of Public Roads in the late 1920s for the design of secondary road construction, based mainly on grain size, plasticity and shrinkage characteristics. The system contained too many shortcomings for good highway design and many variants followed. One such modification by Steele (1946) was adopted for the *AASHO Soil Classification System.*

buried channel The former course of a river, marked by a channel that is plugged by alluvium. Many rivers possess deep former channels that were cut when sea level was lowered during glacial episodes, and have subsequently become aggraded as sea level has risen. The present river channel may or may not be along the line of such a channel. Buried channels are normally *aquifers* supplied by the ground surface, and have proved hazardous during the construction of tunnels beneath the present-day levels.

button bits Used in downhole percussion air drilling, these bits have replaceable tungsten carbide buttons set proud of the face and rely on heavy impact to shatter the rock. Drilling rates are generally faster than with cross-bits. *See rotary percussion drilling.*

byte Portion of a digital tape word comprising a number of *bits* forming a unit.

C

^{14}C *See carbon-14 (^{14}C).*

cable-tool drilling A method of drilling by lifting and dropping a string of tools suspended on a cable. The string comprises a drilling bit to break up the formation, attached to a drill stem with jars to provide weight on the bit and keep the hole vertical, and a swivel joint to allow of rotation of the string. The cable is arranged to be taut at the moment of impact between bit and formation, which, owing to the lay of the cable, causes the bit to rotate as the cable alternatively stretches and contracts. This aids the cutting action of the bit and helps keep the borehole straight and circular. Also known as *percussion drilling*, or *standard* or '*yo-yo*' *drilling*, this method has been dated back to 600 BC for brine drilling in China. High-capacity rigs can currently drill to 1500 m or more (maximum depth of 3397 m recorded at New York in 1953). Main advantages over other types of drilling are simplicity, low cost, limited operating skill required, ease of repair, limited water requirement and ability to drill through fractured, broken or cavernous rocks. Disadvantages include limitation on drilling rate, limited lack of control over borehole stability and flow from permeable strata, and generally frequent drill string failures. *See also rotary percussion drilling.*

caisson A structure which is sunk through ground or water for the purpose of excavating and placing work at the prescribed depth and which subsequently becomes an integral part of the permanent work. A *box-caisson* is a caisson closed at the bottom but open to the atmosphere at the top. An *open caisson* is a caisson open at the top and the bottom. A *compressed-air or pneumatic caisson* is a

42

caisson with a working chamber in which the air is maintained above atmospheric pressure to prevent the entry of water and ground into the excavation.
Reference: CP2004 : 1972.

calc-alkali rock A term applied to those igneous rocks in which there is a substantial proportion of calcium ions, as opposed to sodium or potassium ions. Thus, the feldspar in a calc-alkali rock tends to be calcic plagioclase, and calcium-bearing ferromagnesian minerals such as pyroxenes and amphiboles may also be present.

calcareous Containing calcium or its compounds. In sedimentary usage, the term usually implies the presence of calcium carbonate, while in metamorphic petrology the implication is that calcium silicates derived from original carbonates are present. Calcareous is not used in igneous petrology, the term 'calc-alkali' being used in its place.

calcrete The hard capping of 'caprock' formed at ground level as a result of the upward leaching of limestone rock due to the high evaporation rate in dry arid climates. This causes a net upward movement of groundwater near the surface, allowing soluble minerals and carbonates to be carried into the surface layers, which makes them stronger and denser.

caldera An eruptive volcanic basin of substantial proportions.

Caldwell drill A large mechanical bucket-auger type *rotary drilling* machine used for the installation of large-diameter bored piles.

Caledonian orogeny The phase of mountain building that occurred between approximately 500 and 400 Ma ago, which led to the production of the fold mountain belt that runs from north Norway through Britain and Greenland to North America. *Plate tectonic* models suggest that this fold belt originated during the collision of two former continents, now represented by the shield areas of the Baltic and Canada.

Caledonides The fold mountain belt produced during the Caledonian orogeny. Strictly, the name applies to the part of the belt that lies in the British Isles, but the term is now used as a general term for the whole belt.

caliche Soft clayey limestone.

California Bearing Ratio (CBR) Test A constant rate of penetration test for evaluation of subgrade strengths for road and airfield pavement design. It is expressed as a percentage of the value of 100 obtained for crushed-stone base widely used in California.

Californian bit A heavy-duty percussion drilling bit with sloping shoulders.

caliper log A tool used in *well logging* for measuring the variation in cross-section of a *borehole* with depth.

calyx boring, or shot drilling A method of boring using steel shot as the cutting medium.

43

cambering A valley-side phenomenon whereby a bed of hard caprock overlying a softer bed acquires a secondary dip into the valley as a result of movements in the underlying weaker stratum. Cambering is particularly found in valleys that have been subject to periglacial activity—for example, around the Jurassic strata of the English Midlands. In this case the harder beds of limestone have become cambered as a result of movements in the interbedded clays. Cambering is associated with *valley bulging* (the deformation in the softer strata) and with the formation of *gulls* (open cracks between individual cambered blocks).

Cambridge sampler A *drop sampler (corer)* used for obtaining samples from the sea-bed.

Camkometer A self-boring pressure meter used to determine strength and deformation characteristics of soil. Developed at Cambridge University, England, it is drilled into the ground by rotating a cutter at the base of the instrument and simultaneously applying a flushing fluid and pushing the instrument against the bottom of the hole. The pressure meter is thus tight against the soil, which is not disturbed and maintains the same *in situ* stress conditions. In clays soil parameters including *in situ* total lateral stress, *in situ* shear modulus, undrained shear strength and sensitivity can be determined, while in sands the angle of shearing resistance with respect to effective stress and the shear modulus can be found. *See also pressure meter test.*

CANDE A computer program acronym, standing for Culvert ANalysis and DEsign, developed for the structural design, analysis and evaluation of buried culverts but also used for a variety of other soil–structure interaction problems.

cap, capping piece or distributor A short length of timber placed over a joint where two *walings* abut, to take the thrust of a *strut*.

capillarity The phenomenon of surface and interfacial tension at the interfaces of solids and liquids, characterised by the movement of liquid along a small-bore tube (capillary tube) when placed in the liquid. The movement *(h)* may be in either direction, dependent upon whether or not the capillary walls have an affinity for the liquid, and is equal to

$$\pm 2T\frac{\cos\alpha}{r\gamma g}$$

where h = the height of the capillary movement; T = the surface tension; α = the angle of wetting (zero for most soil mineral–water systems); r = the radius of the capillary tube; γ = the density of the liquid; and g = gravitational acceleration. For circular capillary tubes and taking $\gamma = 1$ for water,

44

$$h = \frac{ST}{ng}$$

where S = the internal or capillary surface per unit volume; and n = the pore space per unit volume.

capillary action   The movement of a fluid through small voids or fine-bore tubes due to surface tension of the fluid.

capillary fringe   The zone immediately above the water table where groundwater is drawn up by surface tension to fill all or some of the interstices in the ground. The water is at a pressure less than atmospheric and is continuous with the water below the *water table*.

capillary head   The difference in head between the meniscus in a capillary opening and the fluid with which the opening is in contact.

capillary interstice   An opening or void in contact with a fluid which is small enough to allow surface tension to produce a capillary movement of the fluid.

carat   (1) The standard unit of weight for diamonds and other precious stones; equal to 0.2 g (3.086 grains). (2) The standard of fineness for gold; pure gold equals 24 carats.

carbon-14 (^{14}C)   A radioisotope of carbon with a half-life of 5730 a. It is produced by the reaction between nitrogen and neutrons produced by cosmic rays in the upper atmosphere.

carbon-14 dating   A method of age determination based on the quantity of *carbon-14* (radiocarbon) in organic material. Living organisms incorporate ^{14}C into their tissues in a proportion that is controlled by the level of atmospheric ^{14}C. At death the uptake of ^{14}C ceases, and the proportion in the tissues then decreases steadily owing to radioactive decay. Measurement of this proportion is thus a measure of the age of the material. Because of the relatively short half-life of radiocarbon, the method is restricted to material less than about 70 000 years in age. It is particularly used for the age range 20 000 years to the present. Since the original ^{14}C level depends on the atmospheric production of the isotope, corrections are made to allow for changes in the rate of production in the period in question. It may be noted that substantial errors can be introduced by contamination, especially as a result of the introduction of small amounts of modern carbon into old materials.

carbonado   The toughest form of industrial diamond. It is usually black or dark grey and has no normal cleavage planes.

carbon dioxide (CO_2)   Gas derived from the decay of organic material and present in surface and subsurface water supplies. CO_2 ionises in water, forming carbonic acid, and may contribute to some types of corrosion.

Carlson stress meter An earth pressure cell for the measurement of boundary stress conditions, in which a thin film of mercury, when pressurised, deflects a diaphragm, thereby changing the output of a strain meter unit.

cartographic plotter An instrument for producing maps and plans from stereoscopic pairs of aerial photographs. A pantograph system is incorporated which allows of both plotting and heighting (contouring) to scale by following the photographic information with an optical system.

Casagrande liquid limit apparatus Apparatus essentially comprising a cup which by a hand-cranked cam can be made to repeatedly drop onto a spread pad and into which soil at varying consistencies is spread. A standard groove is formed in the soil pat and the number of 'blows' required to cause the two halves of the soil in the cup to close is recorded. A plot of blow count against soil moisture content is then made, from which the *liquid limit* is given as that moisture content for 25 blows. *See Atterberg limits and soil consistency; cone penetrometer.*

Casagrande Soil Classification System *See soil description and soil classification.*

cased pile A pile with a permanent shell or casing which is subsequently filled with concrete.
Reference: CP 2004 : 1972.

casing Lining tubes used to line a *borehole* to prevent the sides from collapsing.

casing collar The coupling used to join lengths of *casing*, or lining tubes, together.

casing shoe A thick-walled steel band fixed to the base of *casing* to protect it when driving into stiff and/or strong ground.

cataclasis Mechanical breakdown or fragmentation of rock as a result of deformation.

cataclastic rock, texture Rock which is fragmented on a small scale. *See breccia.*

catalyst A substance whose presence allows a chemical reaction to take place while remaining unaffected itself.

cathode The *electrode* through which direct current leaves an *electrolyte.*

cathodic protection A means of providing immunity to metal from corrosive attack in underwater or underground structures such as jetties or pipelines by causing direct current to flow from its electrolytic environment into the metal. Either *galvanic action* using sacrificial anodes may be used or an *'impressed current'* system adopted whereby an external source of electric current is provided.

cation An ion carrying a positive charge of electricity by virtue of which it is attracted on electrolysis to the *cathode.*

cation exchange The property that allows ions (cations or anions) of one element to be absorbed at the surface of a colloid and allows of their replacement by ions of another element.

cavitation The phenomenon resulting from the pressure in a liquid becoming less than the hydrostatic pressure, allowing of the release of air and generating shock waves. Can cause severe corrosion of metal parts.

cement A powder artificially formed by fusing limestone clay or shale with iron oxide. On mixing with water the paste sets hard. Natural cementation of rock strata can take place by percolation of calcium minerals.

cement (mineral) The material which in a sedimentary rock binds together the separate mineral particles. A variety of materials may act as a cement, the most common being calcium carbonate, silica, iron oxides and clays. The resistance of a sedimentary rock to weathering and erosion is largely determined by the nature of the cement, as is the durability of the rock when used as a construction material. In general, silica cements are the most durable, carbonates and oxides less so, and clays least of all. The origin of mineral cements is not fully understood; however, the principal causes are deposition from circulating groundwaters, solution and redeposition of primary sedimentary material, including mineral grains and biogenic remains, and secondary oxidation of iron compounds.

cell, electric A complete electrolytic system consisting of a *cathode*, an *anode* and an intervening *electrolyte*.

cellular cofferdam A cofferdam enclosed by a wall consisting of a series of filled cells of circular or other shape in plan. It is normally formed by driving steel sheet piling in configurations such as those shown in *Figure C.1*. It may also be used to retain soil and form permanent quays or wharfs. *See* **cofferdam** and **double-wall cofferdam**.

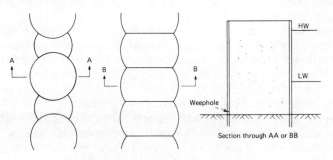

Figure C.1 Types of cellular cofferdam

47

Cemcore auger-injected pile A piling technique introduced by Cementation Piling and Foundations Ltd, UK, whereby a continuous-flight auger is screwed into the ground to the required depth and sand-cement grout pumped under pressure down the hollow stem as the auger is withdrawn. *See also* **Concore auger-injected pile.**

cement-bond log A well log showing the amplitude of the acoustic wave generated by a sonic source that gives an indication of the degree of bonding obtained between cement and the *casing*/formation.

centipoise A unit of viscosity (symbol, cP). $1 \text{ cP} = 10^{-3} \text{ Ns m}^{-2}$. Water at $20°C$ has a viscosity of 1.005 cP.

centraliser A device attached to the rods of a drill string or those used in down-hole testing (e.g. vane and Dutch cone penetration tests) to keep them central and minimise friction and bending.

central processing unit (CPU) A computer term for the centre of a computer's data-handling and calculation processes.

centrifuge Equipment which allows models to be rotated so that the resulting centrifugal forces speed up *consolidation* effects and thereby reduce the time scale required to observe the soil-structure interaction.

chalk A fine-grained limestone composed of organic detritus from planktonic organisms. Chalk differs from other limestones in its purity, which suggests deposition in an area remote from continental sources. The best-known deposits are of Cretaceous age and are widespread in Europe and North America.

channel fill Sediment that has been deposited within the confines of a stream or river channel. The sedimentology of the deposits varies in both the vertical and horizontal directions, owing to the rapid changes that occur in the depositional conditions within the channel.

character log A (wiggle-form) method of displaying the acoustic wave-train, as opposed to the form shown in the *three-D log*.

charge density The number of electrically charged particles or ions per unit area.

chattermark A small-scale feature of surface damage that is formed when a glacier drags a rock fragment across the surface. The markings so produced usually occur as a linear train, similar to those formed when a cutting tool is forced against a hard surface. Chattermarks are one of a number of minor features of glacial erosion: other examples are striae and gouges.

cheesy A description applied to highly colloidal soils which exhibit elasticity when wet and can undergo considerable deformation without rupture but can be broken without much difficulty.

Chemfix process This is a method whereby chemicals are added to liquid effluents and sludges to convert them to earthlike solids. A wide range of toxic liquids can be treated including those from petrochemical, steel, automobile, electronics, chemical and metal finishings industries, and the resultant material can be left as cast or placed as fill and the land used for agricultural or recreational use. The process was developed in the USA and is operated under exclusive licence by Wimpey Laboratories in the UK.

chert A hard, siliceous material that occurs within beds of limestone. Chert may exist as both nodules and thin beds, and is believed to be formed by solution and redeposition of tests of organisms. In many cases chert is indicative of a deep-water environment. Chert nodules that are found in association with chalk are commonly termed *flint*.

china clay The residual product formed by the weathering of alkali feldspars, usually in granite. Chemically, china clay is dominantly the clay mineral kaolinite. Large quantities of china clay have been extracted from the Cornubian granitic batholiths in the west of England.

chisel *See shell and auger boring.*

chi-square test A statistical test in which the known frequency of occurrence of an event is tested against the frequency that is expected on the basis of some hypothesis. The result of the test gives the probability that the observed distribution has arisen by chance, and so provides an assessment of the confidence that can be placed in the original hypothesis.

chlorination The introduction of chlorine solution into a water supply for sterilising and the destruction of nuisance bacteria.

chlorine (demand) The difference between the amount of chlorine supplied to a water supply and that remaining at the end of the contact period. Demand varies, dependent upon the amount of chlorine supplied, contact time, pH and temperature.

chlorine log Obtained by down-hole radiation logging by counting the gamma-rays produced by capture of thermal neutrons by chlorine in the formation. Now mostly replaced by *neutron lifetime log* and *thermal decay time log*.

chlorine (residual) Residual chlorine may be present in a water supply either as free available chlorine (hypochlorous acid and/or hypochlorate ion) or as combined available chlorine (chloramines and other chlorine derivatives). Both types may be present at the same time.

'C' horizon The weathered top of the geological deposits immediately below the soil layer. *See 'A' horizon; 'B' horizon.*

chromium-51 (^{51}Cr) A radioisotope of chromium with a half-life of 27.8 days, used in radio-tracing.

churn drilling *See percussion drilling.*

49

CIPW Classification A classification system for igneous rocks devised by Cross, Iddings, Pirsson and Washington in 1903, in which the percentages of the metal oxides are combined according to a set of rules to yield 'normative' minerals. The rules are based on the reaction chemistry of an igneous melt, and so there is some correspondence between the normative composition and the actual minerals present, but it is not absolute. The normative mineral values are then worked into a system of classes and sub-classes to yield the classification. This classification is not widely used at the present day, although the calculation of normative compositions is of some importance in comparative igneous petrology.

circle (lattice) charts Circle (lattice) charts comprise curves of constant subtended angle between pairs of shore stations which facilitate the positioning of vessels or marker buoys during, for example, a marine investigation. By plotting sextant angles relative to two intersecting sets of curves, the position of the observer can be fixed. Thus, in *Figure C.2*, if sextant angles are measured between the two pairs of shore stations AB and CD, the position of the sighting vessel can be plotted at P. Note that the position of stations B and C can be coincident.

circulation The closed-circuit movement of a drilling fluid (e.g. mud from the mixing pit) down the drill pipe to the drilling bit and back

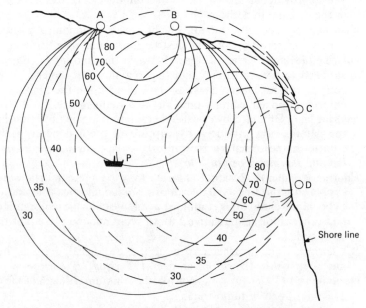

Figure C.2   Circle (lattice) chart. A, B, C, D — shore survey stations

50

up the annulus between drill pipe and borehole to the pit. Where the formation being drilled is porous, some of the drilling fluid may be lost in this material.

cirque A bowl-shaped depression found in mountainous areas, formerly occupied by a small glacier. Also termed a *corrie* (Scotland) or a *cwm* (Wales).

cladding or cleading A timbering term applied to heavy close sheeting used during the excavation and sinking of shafts. It is also applied to the method of covering the sides of buildings with, for example, timber, asbestos or metal sheeting and vertically hung tiling.

claquage grouting A method of grouting involving the injection of grout into planes of weakness of a soil, thereby compacting it. It may also provide a lattice of impermeable membranes, thereby significantly reducing the permeability of the soil mass. The method has limitations where *heave* may be a problem.

classification of landslides Landslides occur in both soils and rocks and are generated by one or more of a variety of causes, including geological and hydrological conditions, weathering effects, climate, topography, *earthquakes* and construction activities by man. A publication by Zaruba and Mencl in 1969 classified landslides according to the shape of the sliding surfaces into '*asequent*' (i.e. those occurring along curved cylindrical failure surfaces), '*consequent*' (those with failure surfaces following planes of weakness such as bedding planes, joints and fissures) and ('*insequent*' (those extending deep into a slope). Skempton (1953) considered the depth-to-length (D/L) ratio of the slide, and Zaruba and Mencl (1969) also classified slides as 'active', 'dormant' or 'stabilised' slides. Hutchinson (1968) classified the mass movements of slopes as: (a) creep movements—include shallow, deep-seated and progressive creep; (b) frozen ground phenomenon—including freeze–thaw movements such as cambering and valley bulging, solifluction sheets, stone streams and rock glaciers; and (c) landslides—including *rotational slips, falls* and subaqueous slides. Skempton and Hutchinson (1975) have also shown that stability hinges on whether the landslide is a first-time slide in previously unsheared ground where the shear strength of the ground is at *peak* or between peak and *residual* values, or whether the slide occurs on pre-existing shears where the strength parameters are at about their *residual* values. The latter type are associated with previous landslides, *colluvium*, periglacial solifluction, tectonic or lateral expansion.

See also: rotational slips; translational slides; slab and block slides; debris flow; mudflows; earthflows; flow slide; solifluction; spontaneous liquefaction; toppling; wedge failure; falls; ravelling.

51

References: *Landslides and their Control*, Zaruba and Mencl, Elsevier, Amsterdam, 1969; 'Soil mechanics in relation to geology', A. W. Skempton, *Proc. Yorkshire Geol. Soc.*, **29**(1), 1953; 'Mass movement', J. N. Hutchinson, *The Encyclopedia of Geomorphology*, ed. R. W. Fairbredge, pp. 688–695, Reinhold, 1968.

classification tests   *See Atterberg limits and soil consistency.*

clay   A natural deposit of fine-grained particles (less than 0.002 mm) consisting mainly of hydrous aluminium silicates derived by the chemical decomposition and mechanical weathering of such silicate minerals as feldspar, pyroxene and amphibole. In engineering terms 'clay' may contain coarser material such as silt and sand, provided that the clay fraction of the whole is not less than 30 per cent. The commonest clay minerals are the montmorillonite, illite and kaolinite groups.

clay auger   *See shell and auger boring*.

clay cutter   A percussive tubular tool used for cutting and removing cohesive soil from a *borehole*.

clay gouge   *See gouge*.

clay minerals   An extensive family of silicate minerals (phyllosilicates) in which silica atoms in tetrahedral co-ordination are arranged in two- or three-layer sheets. The intrasheet locations are filled with various cations—in particular, iron, magnesium, calcium and potassium—and, in addition, further cations may be adsorbed onto the crystal surfaces. (*See cation exchange*.) Clay minerals also contain hydroxyl anions in their lattices, and can hold water molecules in a layer around the crystals. Some minerals (the expanding-lattice minerals) have the ability to increase the separation of the silicate layers and so to increase the amount of adsorbed water. The plastic properties of the clay minerals are closely related to the degree of possible movement between the layers, the most plastic having the greatest movement. In general, three-layer clays are more plastic than two-layer, but anomalies occur due to the ability of some large cations to bind the sheets together, so preventing swelling. Clay minerals are commonly weathering products of other minerals, usually formed by hydrolysis.

claypan   A layer of hard uncemented relatively impervious clay. *See hardpan*.

clay shale   Shale with a greater than usual proportion of fine particles, often formed from sedimentary clays that have undergone very low-grade metamorphism. Clay shales break down to give clays on weathering.

claystone   Clay which has become hardened or *indurated*. Claystones are distinct from clay shales in having fewer parting surfaces.

The term is of restricted use, the term 'mudstone' often being preferred for fine-grained, indurated rocks.

cleat   A stop fixed to prevent movement of a *strut* or *waling* in a timbered excavation.

cleavage   The ability of a rock or mineral to split in certain preferred directions. In the case of a mineral the cleavage results from the configuration of the crystal lattice, whereas in the case of a rock cleavage is more often the result of an alignment or sorting of mineral grains—for example, sedimentary bedding planes or metamorphic foliation.

climatic cycle   A repeating period of time during which a similar number and sequence of wet and dry years are experienced.

CLIMEST   A quick-reply climatological service developed to assist tendering and planning, especially in estimating the time likely to be lost because of adverse weather, provided by Meteorological Office at Bracknell, UK.

clinometer   *See Abney level, or clinometer.*

close sheeting, or close timbering   The placing of vertical or horizontal boards in close formation in order that loose soil formations (especially sand and gravel) may be supported in the face of an excavation.

cobble   A rounded to sub-angular rock fragment ranging in size from 60 mm to 200 mm. *See particle size.*

COBOL   Computer language acronym for COmmon Business Oriented Language.

coefficient of active earth pressure
See active earth pressure; earth pressure.

coefficient of consolidation *(C_v)*   The coefficient of consolidation is given by $k/(m_v \gamma_w)$, where $k = $ *permeability*; $m_v = $ the *coefficient of volume change or compressibility*; $\gamma_w = $ the density of water. *See settlement.*

coefficient of friction   The maximum ratio between shear and normal stress at the point of contact between two solid bodies.

coefficient of passive earth pressure   *See passive earth pressure; earth pressure.*

coefficient of permeability, or hydraulic conductivity   The ratio between the discharge velocity and the corresponding *hydraulic gradient. See permeability.*

coefficient of storage   A hydrological parameter of an *aquifer*, being the volume of water released from storage, or taken into storage, per unit of surface area of the aquifer, per unit change in head.

coefficient of transmissibility   A hydrological parameter of an *aquifer*, being the rate at which water will flow through a vertical strip of the aquifer of unit width and extending to the full saturated thickness, under a *hydraulic gradient* of 1.00. It is equivalent to the saturated thickness of the aquifer multiplied by the *permeability*, where the latter is constant with depth.

coefficient of uniformity The ratio between the sieve size through which 60 per cent of a material passes and the sieve size through which 10 per cent passes. It indicates the grading of the soil—e.g. a closely graded soil has a coefficient of uniformity of less than 3.

coefficient of variation The ratio between the standard deviation of a series of values and its mean.

coefficient of viscosity The *shear stress* required to maintain a unit difference in velocity between two parallel layers of a fluid a unit distance apart.

coefficient of volume change, or compressibility *(m_v)* The ratio, in one-dimensional *consolidation*, between change of volume per unit volume and the corresponding change of effective normal stress:

$$m_v = \frac{e_0 - e}{\{(1 + e_0)\,\Delta\sigma'\}}$$

where e_0 = initial *void ratio*; e = final *void ratio*; $\Delta\sigma'$ = change in effective normal stress.

cofferdam A structure, usually temporary, built for the purpose of excluding water or soil sufficiently to permit construction to proceed without excessive pumping and to support the surrounding ground. A land cofferdam is one constructed from a land surface; a water cofferdam is one constructed in open water. *See cellular cofferdam* and *double-wall cofferdam*.

cohesion The force which binds and holds together soil minerals by molecular attraction.

Cohron shear graph A type of hand penetrometer used for measuring the *shear strength* of soil and the shearing resistance between soil and metal or soil and rubber. Similar in principle to the *ring shear apparatus*, it measures shearing resistance by pushing either a metal or rubber shear head into the soil and twisting a recording drum until the soil fails. *See also Bromhead ring shear*.

cold mix recycling A process in which reclaimed asphalt pavement materials, reclaimed aggregate materials, or both, are combined with new asphalt, and/or recycling agents in place, or at a central plant, to produce cold-mix base mixtures. An asphalt surface course is required. *See hot mix recycling; surface recycling*.
Reference: *Civil Engineering*, January 1981.

collapse structures Structures that result from the removal of support, followed by movement. They are most frequently demonstrated by sediments: for example, in materials that have accumulated on unstable slopes, or glacial materials that show disturbance due to the melting of contiguous ice bodies. They are also (less frequently) seen in igneous crystal cumulates that have undergone gravity transport within a magma chamber.

collapsing soils Collapsing soils include loess (weakly cemented silt with a loose structure) and some weakly cemented sands, which can collapse when wetted, owing to the cemented structure being destroyed.

collector well A method of water *well* design wherein horizontal screen pipes or galleries constructed into an *aquifer* radiate out from a central *caisson*, typically 4 m inside diameter and 25–50 m deep *(Figure C.3)*.

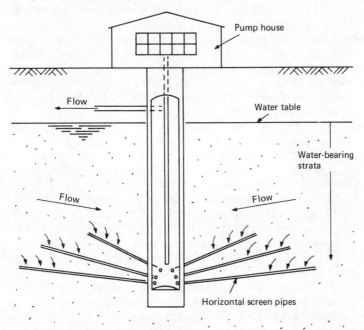

Figure C.3 Collector well

colliery shale Spent waste granular material from coal-mine tips. Satisfactory as filling for road bases and other civil engineering structures, provided that it has been well burnt and that its swelling potential has been fully investigated. Termed 'blaise' in Scotland. Care must be taken to ensure that sulphate content is low enough to avoid chemical attack of any concrete which it may abut.

colloid A substance comprising very small particles (0.05–0.005 μm diameter). In drilling fluids these are mainly dispersions.

colloidal state   The uniform distribution in a fine state of division of one substance (the *colloid*) in another (the *dispersing agent*).

colluvium   Weathered material which has been transported by gravitational forces such as in scree slopes.

colour index   The percentage of dark minerals within a rock. Thus, rocks may be leucocratic (0–30 per cent), mesocratic (31–60 per cent) or melanocratic (over 60 per cent).

commercial vehicle   A commercial vehicle is defined for the purposes of road design as a public service vehicle or a goods vehicle having an unladen weight of more than 15 kN.

compact   A term applied to granular soils which require the use of a pick for excavation and which have a *relative compaction* greater than 90 per cent.

compaction   The process of causing soil particles to pack more closely together by rolling or other mechanical means, thereby increasing the *dry density* of the soil.

compaction factor   The ratio between the increase in density obtained by *compaction* and the density existing before compaction. Usually expressed as a percentage.

compaction grouting   A method of densifying soil and thus strengthening and making it less permeable by injecting into it an expanding bulb of highly viscous grout. This action compacts the soil by moving the soil particles closer together. Compaction grouting has been used as a remedial measure for lifting the foundations of light structures subject to settlement problems.

compass   An instrument for measurement of direction with reference to the magnetic meridian, comprising a magnetised needle, a graduated circle and a line of sight. The main types are the surveying compass, circumferentor or miner's dial and the prismatic compass. The latter is a lighter and simpler hand-held instrument incorporating an optical device that allows the observer to read the required bearing while sighting the instrument.

compensated foundation   *See floating foundation.*

competent bed   A bed is said to be competent if it is stronger than those around it and thus controls their deformation. In a mixed sequence of folded beds the competent layers will be folded uniformly with little change of thickness, whereas the surrounding *incompetent beds* will display thickness changes due to differential loading during folding. It should be noted that competency is a relative term, depending on the environmental conditions during folding and the material contrasts within a particular sequence.

complex phosphates   Complex phosphates such as sodium tetra-phosphate and sodium acid pyrophosphate are very effective degelling agents when added to fresh-water bentonite suspensions,

56

producing reduced viscosity and thus allowing of sedimentation of the suspended solids.

composite Franki *See Franki piling systems.*

compressed-air or pneumatic caisson *See caisson.*

compressibility The decrease in thickness per unit thickness per unit pressure increase. *See modulus of compressibility.*

compression The time-independent volume change associated with an increase of effective stress in a soil. Compression should be distinguished from *consolidation*, which is the time-dependent expulsion of air or water necessary to achieve the compression.

compressional wave An elastic body wave such as a *p-wave*, dilational wave or longitudinal wave in which the motion of the particles is in the direction of propagation. The velocity of these waves is

$$Vp = \sqrt{\frac{E(1 - \mu)}{\gamma(1 - 2\mu)(1 + \mu)}}$$

where μ = Poisson's ratio = $\dfrac{\frac{1}{2}(Vp/Vs)^2 - 1}{(Vp/Vs) - 1}$

and E = Young's modulus of elasticity; γ = the mass density = density/g; g = gravitational acceleration; Vs = velocity of *transverse wave*.

Reference: *Geophysical Prospecting for Oil*, 3rd edn, Milton B. Dobrin, McGraw-Hill, 1976.

compression index (C_c) The slope of the virgin compression curve in a semilogarithmic plot of effective pressure against *void ratio*, obtained from a *consolidation* test on a soil sample. *See settlement.*

compressive strength *See unconfined compression test.*

Compton effect The partial loss of energy from a photon to a free electron resulting from interaction between them.

computer Electronic equipment for analysing data in accordance with a predetermined program. Comprises essentially an input stage for receiving the program and data to be worked on, a memory for storage of this information, a processing unit to undertake the analysis and an output section which can be in the form of a visual display or a typed copy, or both.

computer language The form in which plain-language data and programming instructions are put, in order that a computer can undertake an analysis. Computer languages include *ALGOL, CANDE, COBOL, PAFEC, BASIC, FEMALE, PL-1*, various versions of *FORTRAN* and others.

conchoidal fracture A pattern of fracture in which the broken surface is curved rather than planar. This occurs when the material lacks a well-defined crystal lattice, and so is characteristic of vitreous or amorphous materials.

Concore auger-injected pile A piling technique introduced by Cementation Piling & Foundations Ltd, UK, whereby a continuous flight auger is screwed into the ground to the required depth and concrete is introduced down the hollow stem as the auger is withdrawn. *See also* **Cemcore auger-injected pile.**

conductance *Specific conductance* is inversely proportional to electrical resistance and is a measure of the ability of a liquid to allow of the passage of an electric current through it. It is related to the concentration of ionisable salts in the liquid.

conduction The transmission of heat from particle to particle of a substance.

conductor, electrical A substance (mainly metal or carbon) in which electric current flows by movement of electrons.

conductor pipe A short length of *casing* driven through *overburden* at the top of a *borehole*, or leading from the deck of a drilling platform down to the sea or river bottom, to act as a guide for the main drill string and protect it from wave action, etc.

cone bits *See rolling-cutter bits.*

cone index (CI) *See WES mobility cone penetrometer test.*

cone index gradient (CIG) *See airfield cone penetrometer.*

cone penetration test (CPT) A defined term to include tests variously called *static penetration test, quasi-static penetration test* and *Dutch sounding test.*

Reference: *Proc. 9th Int. Conf. on Soil Mech. and Found. Eng.*, **3**, Appendix A, p. 99, Tokyo, 1977.

cone penetrometer The preferred apparatus for determining the *liquid limit* of a soil in which a 30° angle cone of polished stainless steel or duralumin, 35 mm long and having a mass of 80 g, is allowed to penetrate the soil, which is mixed with water to a range of moisture contents. The penetrations are plotted against moisture content on a graph and the liquid limit is taken as the moisture content corresponding to a cone penetration of 20 mm.

cone of water-table depression The conically shaped surface of the water table in an unconfined *aquifer* surrounding a pumped well. *See Figure C.3.*

confined groundwater A body of water overlain by relatively impervious material which effectively severs hydraulic continuity with the overlying groundwater. When penetrated by a well, water rises up the well to above the top of the water body.

conglomerate A general name for cemented sedimentary rocks whose particles are of *gravel* size or larger.

connate water Water entrapped in the interstices of sedimentary strata at the time of deposition.

consistency (1) A measure of the shear strength of cohesive soil. *See shear strength; soil strength.* (2) The index of grout state. Measured by pendulum-type viscometer or other suitable instrument.

consistency index (CI) Also called the *relative consistency (C_r)*. *See Atterberg limits and soil consistency.*

consolidated-undrained test *See shear strength. See also triaxial compression machine.*

consolidation The process whereby the application of pressure on a soil layer over a period of time causes a reduction in its volume by expelling fluid from the pores and by packing the soil grains closer together. The amount and rate of consolidation are important factors in the computation of *settlement* of structures and thus their foundation design. These parameters can be determined in the laboratory with a consolidation machine or *oedometer*. Three stages of soil deformation under load are recognised: (a) an immediate stage due to elastic deformation of the soil, (b) primary consolidation due to the slow expulsion of pore water and (c) secondary consolidation following full dissipation of the excess pore-water pressure. The latter stage is not generally of importance except in certain cases such as those involving organic soils, where significant consolidation can be experienced.

consolidation settlement *See settlement.*

constant-head permeability test *See pumping-in permeability test.*

constant rate of penetration test (CRP test) A test carried out to determine *ultimate bearing capacity* of a soil, whereby a plate is pushed into the ground at a constant rate of penetration. Marsland describes such tests using plates of 38–868 mm diameter with penetration rate of 2.5 mm/min. The ultimate bearing capacity is taken as the pressure reached when the settlement is equal to 15 per cent of the plate diameter, if not otherwise clearly defined by the test. *See also incremental loading plate bearing test.*

Reference: 'Large *in situ* tests to measure the properties of stiff fissured clays', A. Marsland, *1st Aust.–N. Zealand Conf. on Geomechanics*, Melbourne, 1971.

constrained modulus The reciprocal of *coefficient of volume change, or compressibility (m_v)*. *See also stress .strain relationships.*

contact grouting Contact grouting usually refers to grouting up the annulus between the concrete lining of a tunnel and the surrounding ground. The grout holes are arranged in rings around the tunnel, grouting being started at one end of the tunnel and advancing continuously towards the other. The lower holes in each ring are normally grouted first and the uppermost last.

contact stress transducer An *earth pressure cell* for measuring normal and shear stresses, of particular importance in embankments and soil–structure interfaces on, for example, *piles* and *retaining walls*.

contamination Reduction in water quality by chemical or bacterial pollution to a point where a health hazard exists.

contiguous bored piles A system of *retaining wall* construction whereby *bored piles* are installed in a single or double row and so positioned that they touch or are very close to each other. Gaps between piles which may allow seepage in granular soil can be grouted up. *See also* **secant (interlocking) piles**.

continuous piezometric logging A method allowing of measurement of positive pore pressures at any depth in soils or rock penetrated by a *borehole*. The sequence of operation comprises (a) placing perforated *casing* in borehole and filling with grout; (b) redrilling to remove grout from within casing; (c) inserting thin membrane within casing plus high-pressure cap to borehole if high pore pressures are to be measured; (d) lowering special 'torpedo' to required test level, comprising central section with inflatable air-bag and inflatable packers at top and bottom of torpedo; and (e) conducting test by inflating central air-bag to expel water from test section, inflating packers, followed by deflation of air-bag, allowing external pore pressures in surrounding ground to equilibrate with test section pressures. Pressures are measured by a *transducer*, and the rate at which they equalise allows *permeability* to be estimated, or twin-tube *hydraulic piezometers* can be used to measure permeability directly. *See also* **piezometer**.

continuous seismic reflection profiling A technique which produces, along the track of a survey vessel, a continuous printed record of the sea-bed and sub-sea-bed discontinuities. Acoustic pulses of short time duration, reflected from the sea-bed and underlying substrata, are detected by a synchronised recording device which displays the seismic trace as a sub-bottom geological profile showing the two-way travel times to different reflective horizons. The type of equipment used to provide the acoustic energy depends on the survey objectives, and may be *airguns* (air discharge); *sparkers* (electrical discharge); *gas guns* (chemical combustion); *boomers* (electromagnetic induction); or *pingers* (piezoelectrical or magnetostrictive). In general, pingers and high-resolution boomers are suitable for resolving near-surface layering; standard boomers and sparkers for coarser and thicker overburden; and airguns for high-resolution shallow-water profiling.
Reference: *Sea Surveying*, ed. A. E. Ingham, Wiley, 1975.

continuous trenching machine An excavator having an arm fitted with a continuous chain to which buckets of the appropriate 'trench-width' size are attached. It is effective in suitably consistent ground conditions but unable to cater for obstructions that may be in its path.

continuous-velocity log   *See sonic log.*

contract   A contract is defined in the British ICE Conditions of Contract (5th edn, June 1973) as comprising the conditions of contract, the specification, the drawings, the priced bill of quantities, the tender and its written acceptance, and the contract agreement.

contract drawings   The drawings and plans showing the character and scope of the work to be executed, which are prepared or approved by the *engineer* or *owner* and referred to in the contract documents.

contractor   A contractor is defined in the British ICE Conditions of Contract (5th edn, June 1973) as the person or persons, firm or company whose tender has been accepted by the *employer*, and includes the contractor's personal representatives, successors and permitted assigns.

convolute bedding   Disturbed sedimentary bedding that is usually confined to one horizon. It is formed by secondary disturbance by, for example, downslope movement or the expulsion of pore water.

Cooper–Jacob analysis   A simplification of the *Theis method* for calculating the transmissibility and storage coefficients of an aquifer where the well function $(u) = r^2 S/4Tt$ is very small ($\leqslant 0.01$), and where S = the coefficient of storage; r = the distance from the well to the observation hole; T = the coefficient of transmissibility; and t = time.

Reference: 'Generalised and graphical method for evaluating formation constants and summarising well-field history', H. H. Cooper, Jr and C. E. Jacob, *Trans. Am. Geophys. Union*, **27**, 1946.

core barrel   A steel tube fitted with a *coring bit* at its lower end that can be rotated by a drilling machine to bore into consolidated strata and allow a core of the material to be recovered for identification and testing if required. Four major types have been evolved: *single-tube core barrels*, for use in hard compact formations; *double-tube rigid barrels*, for use in medium hard to hard formations which may be somewhat fractured and in soft formations if relatively compact and unfractured; *double-tube swivel-type core barrels*, for coring friable shattered formations; and *double-tube swivel-type face ejection core barrels*, for coring badly shattered, soft weathered or partly unconsolidated formations. Other barrel types include triple barrels, useful for coring through rock containing thin bands of softer material; and barrels used with wire-line systems, where the core can be recovered by retrieving the core-laden inner barrel through the drill string, the outer barrel and drill string remaining in the hole until the core bit needs to be replaced.

core cutter method for field density determinations *See field density tests.*

core recovery The total length of core recovered from a drill hole, expressed as a percentage of the length of the hole. *See also rock quality.*

coring bit Used in *rotary core drilling*, the *bit* or crown is annular in shape and is screwed onto the lower end of the outer core barrel (sometimes via a reaming shell). Since it is the function of the bit to grind and crush the rock material, pieces of tungsten carbide or industrial diamonds are set in a softer metal matrix around the end of the bit. Other systems are occasionally used but diamond or tungsten carbide set bits are the most common, the latter being used mainly in the weaker rocks. No type of bit has universal application but, in general, the size of diamond to be used varies inversely with the strength and abrasiveness of the rock. Sizes of diamonds in set bits vary from about 2 diamonds per carat to about 100 diamonds per carat and smaller sizes in impregnated bits.

Coriolis effect The effect of the acceleration of a body in motion with respect to the Earth resulting from the Earth's rotation.

corrosion The chemical or electrochemical reaction of a metal with its environment, resulting in its progressive degradation or destruction.

Reference: CP 1021: 1979.

coset A sedimentary unit composed of *cross-beds* which indicate a consistent direction of palaeocurrent flow.

cosmic rays High-energy radiation consisting of charged particles which enter the Earth's atmosphere from outer space.

cosmic water Water introduced to the Earth from space by meteorites.

Coulomb's earth pressure theory Coulomb's classical earth pressure theory for cohesionless soils was based on the concept of failure by a wedge of soil behind a retaining wall rising at an angle from the base of the wall. A plane failure surface was assumed and also that the direction of the thrust on the wall was known. In *Figure C.4(a)*, AB is an arbitrarily chosen failure plane and W is the weight of the failure wedge ABC. The resultant force P_f acts through the centre of gravity of the wedge at an *angle of obliquity* ϕ (the *angle of internal friction* of the soil), and the resultant force on the wall, P, also acts through the centre of gravity of the wedge at an arbitrarily chosen angle of obliquity α. Since the magnitude and direction of W are known and the directions of P and P_f are known, a force diagram (*Figure C.4(b)*) will give the magnitudes of the latter forces. As the obliquity angle α is approximately equal to ϕ, trial and error using a number of failure planes will allow the critical value of the lateral force, P, to be determined. *See also Rankine's earth pressure theory.*

Reference: 'Essai sur une application des règles des maximis et minimis à quelques problèmes de statique relatifs à l'architecture', C. A. Coulomb, *Mém. acad. roy. pres. divers savants*, 7, 1776.

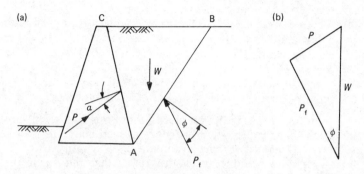

Figure C.4 Coulomb's earth-pressure theory

Coulomb equation for shear strength *See angle of internal friction.*
counterfort drain A stone-filled drain constructed in the side of a cutting to facilitate drainage and thereby increase its stability.
counterfort wall A wall or pier constructed at right angles to a **retaining wall**, usually on the side of the retained material and thus hidden from view, provided to increase its bending strength and stability against overturning.
counterpoising The procedure of placing rock or high-density material at the toe of an embankment to provide an overall increase in the stability of the slope.
country rock Pre-existing rock into which is intruded igneous material. Thus, one speaks of the relationship of the intrusion to the country rock, the degree of alteration of the country rock, etc.
cowbelly A colloquial term for a very silty soil characterised by its spongy feeling and jelly-like behaviour when rolled or vibrated (e.g. by being jumped on). This is due to the induced pore-water pressures being unable to dissipate quickly.
[51]Cr *See chromium-51 (^{51}Cr).*
creep In general terms, creep is the time-dependent deformation of solids under stress. An indication that a material might be prone to creep may be obtained from cyclic loading carried out during compression tests. The time–strain behaviour of many materials which are susceptible to creep is similar and the following types of movement are discernible:

(a) An *instantaneous elastic strain* occurring on application of load.

63

(b) A period of *primary creep*, with rate of strain decreasing logarithmically, due to elastic flow. If the stress is removed during this period, there is first a recovery of the instantaneous elastic strain followed by a time-dependent period of the recovery of the primary creep.
(c) A period of *secondary creep*. During this phase the rate of strain is constant, and if unloaded, a specimen would show permanent deformation. (This is sometimes called *pseudoviscous flow*.)
(d) If the stress is continued, *tertiary creep* may occur with a rapidly increasing rate of strain until failure occurs.

Laboratory experiments on creep-prone soils and rocks have indicated that the 'long-term' strength of such materials may be substantially less than their 'instantaneous' strengths. However, 'long-term' may be on the geological rather than the engineering time scale. Microcracking and flaw development occur during creep with accompanying microseismic activity (detected as acoustic emission). This may occur at less than half instantaneous failure strength.

crib, or grillage   A timbering term for a layer of timber or steel in either one or two directions placed beneath a *shore, sill, or soleplate* for the purpose of spreading the load.
   Reference: CP2004 : 1972.

critical angle   The angle of incidence at which a (seismic) wave will refract along the interface of adjacent media. *See* **headwave**.

critical density   The density at which a cohesionless soil does not suffer volume change during shear.

critical depth of excavation   The depth of an excavation at which plastic flow of the soil into the base of the excavation takes place as a result of the pressure of the surrounding ground. Skempton, 'The bearing capacity of clay', Building Research Congress, London, 1951, gave the critical depth $D_c = N_c(S/\gamma)$, where N_c is a 'bearing capacity' factor dependent on the length, width and depth of the excavation, S is the undrained shear strength of the soil and γ is the soil density. The factor of safety (F) against bottom failure can be assessed from:

$$F = \frac{N_c S}{D(\gamma_1 - \gamma_2) + P}$$

where N_c and S are as given above; $\gamma_1 =$ the density of the soil; γ_2 = the density of the material filling the excavation—i.e. $\gamma =$ the density of water or bentonite mud, as appropriate, or zero if the excavation is dry; and $P =$ the surcharge on the ground surface surrounding the excavation. The factor of safety (F) against **heave** occurring in cohesionless soils is

$$F = 2N_\gamma \left(\frac{\gamma_2}{\gamma_1}\right) K_a \tan \phi$$

where N_γ = the bearing capacity factor of the soil below the excavation; γ_1 = unit weight of soil above the bottom of the excavation; γ_2 = unit weight of soil below the bottom of the excacation; K_a = the coefficient of *active earth pressure*; and ϕ = the internal angle of friction of the soil.

References: *Foundation Engineering Handbook*, H. F. Winterkorn and H. Y. Fang, Van Nostrand Reinhold, 1975; *Soil Mechanics in Engineering Practice*, K. Terzaghi and R. B. Peck, Wiley, 1948.

critical hydraulic gradient A condition where the pressure of an upward seepage of water through a sandy soil becomes equal to the submerged weight of the soil and thus the critical velocity, i, is equal to γ'/γ_w, where γ' = submerged unit weight of soil and γ_w = unit weight of water. As the seepage pressure increases above this point, the sand behaves as a liquid and is accompanied by a violent and visible agitation of the sand particles, a phenomenon commonly known as *boiling*.

critical state The condition in which a soil has reached a *critical void ratio*.

critical state line The graph of *critical void ratio* plotted against the *effective stress* under which that *void ratio* is achieved. The critical state line lies parallel to the *virgin compression line* and slightly below it on a standard void ratio–effective pressure graph. Soils which contract on shearing have a void ratio above critical, whereas soils which expand on shearing have a void ratio less than critical.

critical state soil mechanics A theory of soil behaviour centered on the concept that there is a unique critical state for a given soil that depends on its composition and the level of *effective stress* application. One of the principal postulates is that the critical state may be reached by any combination of stresses—a fact that enables the results of drained and undrained tests to be correlated with each other and with the results of compression tests. The theory has enjoyed several successes, notably the explanation of the prefailure behaviour of normally consolidated soils.

critical velocity The velocity at which flow changes from streamline to turbulent.

critical void ratio The value of the *void ratio* for a particular state of *compaction* of a granular soil below which denser material tends to increase in volume when sheared or vibrated, and above which looser material tends to decrease in volume; thus, at the critical void ratio, the material will neither expand nor contract when disturbed. Sand having a void ratio below the critical value (i.e. in a denser state that at its critical value) is unlikely to suffer sudden *liquefaction* or a *flow slide*.

cross-beds   Primary sedimentary structures which result from the superposition of dune or ripple bed forms and which bear a relationship to the contemporary current direction. In vertical section cross-beds appear as inclined partings or planes organised into larger units or *cosets*. Individual cosets are bounded by erosion surfaces which indicate a fluctuation in stream flow. A complex terminology exists, based on the three-dimensional shape of the coset—hence tabular cross-beds, etc. The terms 'cross-stratification' and 'current bedding' are synonyms.

cross-hole shooting   Methods enabling the time to be measured for body waves to travel between several points at the same depth in a soil or rock mass. Methods include the use of a spark source devised to locate underground cavities and a shear wave technique using a *standard penetration test* hammer as a wave-generating source, the shear waves being measured by receivers in two other boreholes *(Figure C.5)*.

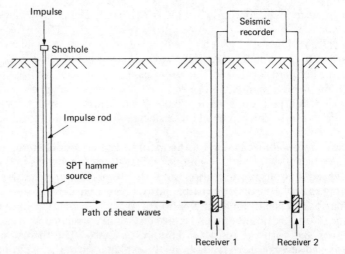

Figure C.5   Cross-hole shear wave technique

cross-section   A diagram showing information across a structure, machine or geological formation, etc., cut by a vertical plane.

crown block   The pulley or sheaves at the top of a drill derrick.

crust   (1) The hardened upper part of an engineering soil profile which is the result of *desiccation* or cementation by precipitated salts. (2) The outermost layer of the Earth, distinguished by a break in the seismic wave profile. There is a distinction between the continental and oceanic crusts on the basis of both thickness and composition. The continental crust is on average 30 km thick and is composed of granitic and granodioritic rocks, whereas the ocean crust is only 10 km thick and is composed of basalt.

crystalline rock   A rock composed of interlocked crystals of primary formation. Thus, the term includes the igneous and metamorphic rocks as well as the sedimentary evaporites.

cryogeology   A general name for those branches of earth science that are concerned with *frozen ground phenomena.*

Culmann line   A curved line obtained by graphical construction, points along which represent values of earth pressure for various arbitrarily chosen failure planes, as discussed under *Coulomb's earth pressure theory.* A tangent to the line drawn parallel to the 'slope line' *(see Figure C.6)* gives the *active earth pressure, P_a,* and the line joining the tangent point with the base of the wall gives the actual surface of sliding.

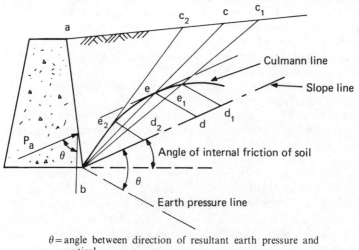

θ = angle between direction of resultant earth pressure and vertical
ed = active earth pressure (P_a), to scale
bc = actual surface of sliding
bd = weight of wedge b c a, to scale

Figure C.6   Culmann line

Culmann's method of slope stability   A method for cohesive soils which assumes failure of a wedge of soil along a plane surface rising at an angle from the toe of a slope *(Figure C.7)*. It gives reasonably accurate results for near-vertical slopes but becomes increasingly

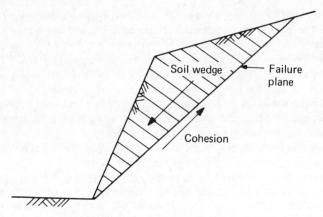

Figure C.7 Culmann's method of slope stability

inaccurate as the angle of slope decreases.

Reference: 'Die graphische Statik', K. Culmann, Zurich: 1866.

cup and vane anemometer *See anemometers.*

curie A measure of the radioactivity of a substance. It is defined as the amount of a radioactive isotope which decays at the rate of 3.7×10^{10} disintegrations per second.

current meter A marine instrument for determining the rate and direction of flow at any depth.

curtain grouting Curtain grouting involves grouting by either the *'stage' method* or the *packer method*, of a continuous vertical curtain of grout to provide a cut-off to water infiltration under, for example, a reservoir dam. Stage grouting involves drilling holes to a limited depth, which are then grouted. The holes are then washed clean, drilled down to a new level and regrouted. Washing out, redrilling and grouting are continued in sequence until the final grouted depth is achieved. In the packer method, holes are drilled to the required final depth, packers inserted at the top of the required grouted zone and the zone grouted up. The overlying zone is then grouted for a specified thickness and the procedure repeated until the total thickness is grouted.

cut-off wall An impermeable barrier provided by a *slurry trench*, *diaphragm wall*, *sheet pile wall*, *grout curtain* or *ground freezing* technique.

cutting, or cut An open-cut permanent excavation provided to allow formation level for a structure (e.g. road, railway) to be below natural ground level. The angle to which side slopes may be formed is dependent upon ground conditions and depth of cutting.

cyanaloc grout   A resin grout, immiscible in water, which forms a very stiff gel when catalysed with sodium bisulphate. It is particularly useful for grouting fractured rock.

cyclic loading   Cyclic loading involves changes in direction of *shear stress* which a material undergoes when subjected to variations in loading similar to that experienced by a soil below the sea-bed due to changes in water pressure caused by wave action, or by stress variations in the ground caused by earthquake waves. Soils suffer a reduction in strength when subjected to cyclic loading, and it has been shown that directional changes of shear stress produce increasing transient and residual pore-water pressures in saturated undrained granular soils. In such soils, where the applied stresses are of sufficient strength and duration, the excess transient pore pressures can build up and approach the confining pressure, with consequent almost complete loss of soil strength. Methods have been evolved for determining the *liquefaction* potential of saturated sands based on a comparison of stresses induced in the ground by earthquake activity with those causing liquefaction in laboratory tests, and also by extrapolating information on the field performance of soils in previous earthquake conditions to similar soils at other potentially active earthquake areas.

References: 'Pore-water pressure changes during soil liquefaction', H. B. Seed *et al.*, *ASCE Jl Geotech. Eng. Div.*, **102**, Pt GT4, pp. 323–346, 1976; 'Sand liquefaction in large scale simple shear tests', P. De Alba, H. B. Seed and C. K. Chan, *ASCE Jl Geotech. Eng. Div.*, **102**, Pt GT9, pp. 909–927, 1976.

cyclic sedimentation   A pattern of sedimentation in which the vertical sequence in a section shows a systematic change, this sequence being repeated several times. The systematic change is usually interpreted in terms of facies replacement, whereas the repetition is interpreted as a change of sea level. A particularly well-known example of cyclic sedimentation (*cyclothem*) occurs in the Carboniferous Coal Measures rocks, in which individual cycles are interpreted in terms of a change from deep to shallow water, and the cyclic change is seen as a periodic change in eustatic sea level.

cyclic shear resistance   The level of cyclic stress required to produce *liquefaction* or a given amount of strain in a specified number of loading cycles. The term replaces the term 'cyclic shear strength'.

Reference: 'Definition of terms related to liquefaction', *ASCE Jl Geotech. Eng. Div.*, **104**, Pt GT9, September 1978.

cyclic strain softening   Cyclic strain softening is defined in relation to *liquefaction* as a stress–strain behaviour under cyclic loading conditions in which the ratio between strains and differential shear stresses increases with each stress or strain cycle. In saturated undrained cohesionless soils cyclic strain softening is caused by increased pore-water pressure. Continued cyclic loading usually leads to increasing axial strains and increasing pore-water pressures, but does not necessarily lead to loss of ultimate shear strength if the material is dilative.

Reference: 'Definition of terms related to liquefaction', *ASCE Jl Geotech. Eng. Div.*, **104**, Pt GT9, September 1978.

cyclothem A multiple repetition of beds of different lithology which are recognisably similar in internal sequence, showing usually minor variations both in thickness and precise sequence of components. *See also* **cyclic sedimentation**.

cylinder An alternative name for an open *caisson* or monolith of cylindrical form.

Reference: CP2004 : 1972.

D

darcy A unit of *permeability*, equal to a flow rate of $1 \text{ cm}^3 \text{ cm}^{-2} \text{ s}^{-1}$ of a fluid having a viscosity of 1 cP under a pressure gradient of 1 atm/cm.

Darcy's law The velocity (v) of flow in a saturated soil equals the coefficient of permeability (k) times the hydraulic gradient (i): $v = ki$.

Dataface system 3 A computer-controlled materials testing system that is indirectly integrated with the testing machine, for recording and analysis of the results. Manufactured by Engineering Laboratory Equipment Ltd, Hemel Hempstead, Herts, UK.

data retrieval system A system of storing data and allowing of their rapid retrieval. Small systems normally involve some form of punched card, individual aspects of information being retrieved by pushing rods into a pack of cards, the arrangement of holes punched in the cards so arranged as to allow certain cards to be lifted free of the pack. Large amounts of information are best stored on magnetic tape in a *computer*, and can then be retrieved very rapidly by suitable programming.

Dawson piling hammer A standard drop-hammer modified by dynamic cushioning to reduce the operating noise level to a minimum.

dead men Forms of *ground anchors*.

dead shore A timbering term for a vertical *strut* supporting a horizontal beam which is carrying a wall or other load from above.

debris flow A type of mass movement which occurs in arid mountainous areas. It consists of poorly sorted granular material which flows at a relatively low water content. Such flows are fast-moving, and tend to occupy consistent tracks at the base of which are built out debris cones or *alluvial fans.*

Decca Navigator System An electronic position fixing system for ships and aircraft. A Decca chain normally comprises four transmitting stations consisting of a Master Station and red, green and purple slave stations operating in pairs—i.e. a Master Station to each slave station. Each pair of stations generates a pattern of stable stationary waves so synchronised that the crests and troughs of the waves are in phase at each transmitting aerial. As a vessel moves between a pair of stations, it crosses the wave pattern and the phase difference between the two signals can be measured by a receiver aboard the vessel. The phase difference between a complete waveform (e.g. peak to peak) is called a *lane,* each lane being identified by a number. The receiver is used to drive 'Decometers', which display the number of the Decca position lines passing through the vessel. The location of the vessel is then fixed by plotting the Decometer readings onto *circle (lattice) charts. See also HI-FIX, MINI-FIX, SEA-FIX.*

decibel Unit used for measurement of the intensity of a single sound relative to some reference level (stated or understood). *See also sound; noise; sound level; equivalent continuous sound level; total sound level; sound attenuation; reflected sound.*

deflectometer A tensioned wire device for measuring ground movement normal to the axis of a borehole. The wire passes over knife edges at a number of points along its length, where any angular deflections are recorded by *transducers. See horizontal movement gauges.*

deflocculate To cause aggregates or flocs of clay minerals in a suspension to disperse. The clay particles are swamped with *cations* which 'coat' the grains and cause mutual repulsion. The technique is normally used as a preparation for *particle size* analysis by a sedimentation method, in which case an agent such as sodium hexametaphosphate (Calgon) is employed.

deflocculation The separation of particles in a colloidal suspension due to their attraction to a dispersing medium or due to the mutual repulsion of such particles with like electric charges.

deformation fabric *See tectonic fabric.*

Degebo friction sleeve *See penetrometer (apparatus).*

degree of consolidation The degree of consolidation at any time (δt) is the *settlement* at that time expressed as a percentage of the total settlement (δc) at the end of consolidation. It is dependent on the applied loading, the drainage conditions and soil *permeability.*

degree of saturation The ratio, expressed as a percentage, between the volume of water and the total volume of voids.

delay time In geophysics, the time taken for a seismic wave to travel from source to receiver, minus the time taken to travel the same horizontal distance at the fastest velocity in the circuit, is the sum of all the delay times in the circuit. Thus, the delay time measures the retardation due to slower layers, and from a study of delay times at different receivers (for the same source) one may deduce the geometry of the subsurface beds.

Delft continuous sampler A sampler developed by Laboratorium voor Grondmechanica, of Delft, Holland, to give a continuous sample of the ground of either 29 mm or 66 mm diameter. The 29 mm sampler is pushed into the ground with a conventional 2 Mg Dutch deep sounding machine. As the sampler is advanced, the sample is fed automatically into a nylon stockinette sleeve which is made impervious by a vulcanising process. The sample within its impervious sleeve is maintained in its original position and dimension by a tension cable fixed to the top of the stockinette sheath and a bentonite–barites supporting fluid of similar density to that of the ground. The 66 mm sample is used in conjunction with the 17.5 Mg Dutch deep sounding machine. The sample within its sheath is fed into a plastics tube filled to the appropriate level with supporting fluid. Both samplers have a depth capability of about 18 m.

Delft electric cone/friction sleeve *See penetrometer (apparatus).*

deliquescence The property of absorbing moisture to such an extent that a substance dissolves in it and becomes a liquid.

delta pile A pile similar to a *vibro pile* but using a rammer inside the tube to compact the concrete.

Denison sampler A sampler for obtaining 60 mm diameter undisturbed samples of hardclays, hardpan, dense sand, cemented soils and other difficult ground. It comprises an outer barrel fitted with a cutting bit and an inner barrel which protrudes below the cutter to ensure that the sample is recovered from undisturbed ground and is unaffected by groundwater or drilling fluid. As the sampler is forced down into the ground, the sample pushes through a core retainer and up into the inner barrel containing a thin wall liner which serves as a permanent sample tube after withdrawal and dismantling of the sampler. The sampler is operated by rotating the outer barrel into the soil in either a cased hole or one stabilised with drilling mud. The inner barrel does not rotate. *See also Pitcher sampler.*

densimeter An instrument for measuring density.

density index The density index is defined as

$$\frac{e_{max.} - e}{e_{max.} - e_{min.}}$$

where $e_{max.}$ = the void ratio in the loosest state; $e_{min.}$ = the void ratio in the densest state; e = the void ratio in the natural state. *See also relative density.*

density log A borehole or well record which records the density of the strata penetrated as determined by using a logging tool comprising a gamma-ray source. *See also **gamma–gamma log, or density log; gamma-ray density gauge.***

density–moisture content relationship The relationship between the dry density and the moisture content of a soil when a particular amount of compactive effort is applied (*see Figure A.3*). *See also **maximum dry density**; **optimum moisture content.***

depth of cut-off The depth reached by the sheet piling or cofferdam walls below excavation level.
Reference: CP2004: 1972.

depth factor, Fox A reduction factor proposed by Fox to be applied to the estimated settlement of a flexible foundation due to the depth of its embedment in the ground.
Reference: 'The mean elastic settlement of a uniformly loaded area at a depth below the ground surface', E. N. Fox, *Proc. 2nd Int. Conf. on Soil Mech. and Found. Eng.*, **1**, p. 129, Rotterdam, 1948.

descriptions of soils and rocks *See **soil description and soil classification.***

desiccation The process of drying of a soil caused by natural means (e.g. lack of rainfall) or artificial means (e.g. furnace or boiler-house).

desk study The first stage in a site investigation, wherein all available information is studied in order to select the site location, if a choice is available, and to base the extent of the proposed field work. Information studied may include land survey, permitted use and restrictions, access, ground conditions, available construction materials, water supply and other utilities.

de-stressed zone When an excavation is made into rock in which in situ stresses exist, these stresses will be removed around the excavation owing to the removal of support. This leads to the formation of a de-stressed zone in which the original stresses have been lowered. The extent of the zone reflects the properties of the rock, the shape of the excavation and the state of the original stress field. Problems may arise as a result of de-stressing, since previously closed fissures may open and so allow of ingress of water or the collapse of blocks into the void.

detector spread The arrangement of *geophones* used in a particular seismic study. Common arrangements for divil engineering are in-line, triangular and fan.

Deval attrition test A test in which rock aggregate is tumbled at a slow speed without the abrasive charge of steel spheres used in the *Los Angeles abrasion test.*

deviator stress   The additional vertical stress applied to a specimen in a triaxial compression test to shear the specimen after subjecting the specimen to the 'all-round' hydrostatic compressive stress—i.e. after application of pressure to the fluid filling the triaxial cell. *See triaxial compression machine.*

diagenesis   *See lithification.*

diaphragm wall   A plain or reinforced concrete wall constructed *in situ* for either temporary or permanent works by excavating a deep narrow trench using purpose-made tools while introducing bentonite mud to replace the excacated spoil, which is in turn replaced by tremied concrete.

dielectric   An insulator or non-conductor of electricity.

die overshot   A 'fishing' tool used in the oil drilling industry comprising a hardened steel tapered die which is lowered down a borehole to fit over the top of a lost drill pipe. The die is rotated to thread on to the pipe and thus allow of its recovery. *See fish.*

Dietert test   A test, now generally superseded, for the determination of the compaction characteristics of soil. It consists of mixing air-dried soil passing a British Standard No. 7 sieve with various quantities of water and compacting the mixes in a 2 in diameter cylindrical mould with ten blows of a cylindrical mass of 18 lb dropping through a height of 2 in and repeating the process after inverting the mould. The length of the sample is measured after ejecting it from the mould to find its *bulk density* and the whole sample is then dried to determine its *moisture content.*

differential settlement   Variation in *settlement* from point to point within a structure.

diffusivity   The quotient of the thermal conductivity and the heat capacity per unit volume.

dip meter   (1) A portable transistorised instrument comprising a weighted probe which can be lowered down a well or borehole on a cable to measure water levels accurately. (2) A well or borehole logging sonde comprising a number of sensors to allow micro-resistivity readings to be taken from which, with the addition of caliper measurements and wall scratches, the magnitude and azimuth of the formation dip can be determined.

discharge velocity   The rate of discharge per total unit area perpendicular to the direction of flow.

discontinuity   Discontinuity is defined as the general term for any mechanical discontinuity in a rock mass having zero or low strength. It is the collective form for most types of joints, weak bedding planes, weak schistocity planes, weakness zones and faults. The ten parameters selected to describe discontinuities and rock masses are as follows and are defined separately: *aperture of discontinuity; block size resulting from discontinuities; filling of* *discontinuity; number of sets of discontinuities; orientation of discontinuity; persistence of discontinuity; roughness of discontinuity; seepage from discontinuity; spacing of discontinuity; wall strength of discontinuity.*

Reference: International Society for Rock Mechanics. Commission on Standardisation of Laboratory and Field Tests, 1977.

(discontinuous) reaction sequence *See Bowen reaction sequence.*

dispersing agent A deflocculating substance used in sedimentation (mechanical) analysis, such as sodium oxalate for the determination of fine-grained (clay and silt) particles.

dispersion The moving apart of particles in a suspension—i.e. the act of deflocculation.

distance–drawdown relationship *See drawdown; radius of influence. See also Figure D.3.*

diurnal tide Tides which have one high tide and one low tide in a lunar day of approximately 25 hours.

diving (air-line) *See airline.*

dog A metal spike having both ends pointed and bent at right angles to the bar and in the same direction; used for joining large timbers together. *See bitch.*

dolly A cushion of hard wood or other suitable material which transmits the blows of the *piling hammer* to the driving cap or helmet at the head of the pile.

doorstopper cell An instrument for measuring in situ biaxial state of stress, consisting of a *strain gauge* which is glued to the bottom of a borehole of about 60 mm diameter which is subsequently stress relieved by over-coring. Strain can be measured by photoelastic disc or rosette of electrical strain gauges *(Figure D.1)*.

Doppler anemometer *See anemometers.*

Dorry test The Dorry test and the modified Dorry test are aggregate abrasion value tests designed to test the abrasive resistance of rocks and aggregates by pressing the sample against a rotating steel disc.

double-acting piling hammer *See piling hammer.*

double-packer grouting Double-packer grouting involves isolating a specified part of a *borehole* by inserting *packers* at the top and bottom of the length to be grouted. Grout is then introduced under pressure within the two packers. If the base of the hole is the lower level of the grouting zone, the bottom packer can then be omitted and the method is termed *single packer grouting.*

double-roll pedestrian-controlled roller *See vibrating rollers.*

double-tube swivel-type core barrels *See core barrel.*

double-tube swivel-type face ejection core barrels *See core barrel.*

double vibrating rollers *See vibrating rollers.*

double-wall cofferdam A cofferdam enclosed by a wall consisting of two parallel lines of sheeting tied together and with filling between

75

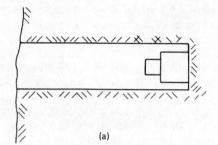

(a)

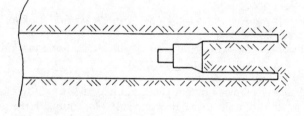

(b)

(a) Strain cell bonded on to end of borehole and strain readings recorded
(b) Borehole extended with diamond coring crown, thereby stress-relieving the core

Figure D.1 Doorstopper cell. Stages in measurement

them, and which is usually self-supporting against external pressure. *See* ***cofferdam*** and ***cellular cofferdam***.
Reference: CP 2004: 1972.

doubly plunging fold A fold whose axis dips away from some centre in two opposing directions. Otherwise termed a *whaleback fold*.

down-hole hammer drilling A *rotary percussive drilling* method using air flush, penetration being obtained by a percussion machine (down-hole hammer) connected directly to the drilling *bit*.

Dowsett Prepakt piles Dowsett Prepakt piles, of 305 mm and 457 mm diameter, are constructed by Dowsett Piling & Foundations Ltd and are formed with a high strength intrusion mortar consisting of fine aggregate, cement and Prepakt's patented intrusion aid. This is designed to ensure a colloidal mix, with minimum water content, which expands on setting. Four types of pile *(see Figure D.2)* are available which are all formed by boring with a hollow centred continuous flight auger:

(a) Where the ground is unstable and when the required depth has been reached, 'intrusion mortar' is pumped under pressure

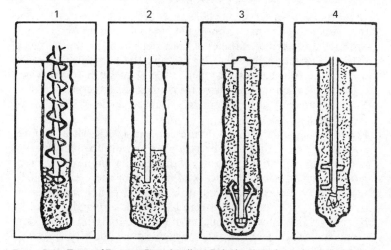

Figure D.2 Types of Dowsett Prepakt pile: 1 Pakt-in-place, 2 cast-in-place, 3 locked-in-place, 4 mixed-in-place

through the stem of the auger, which is slowly removed from the ground until the hole is completely full of mortar, when reinforcement, as required, is installed.

(b) In cohesive soils where the hole is stable and no water problems exist, the hole is bored and spoil removed in one operation. Mortar is then injected into the borehole via a flexible pipe working from the bottom upwards. Reinforcement is placed as soon as the hole is filled with mortar.

(c) Where the pile has to resist uplift forces or bending stresses, the normal reinforcement can be replaced by an anchor assembly which can allow the pile to be post-tensioned. This is known as the 'locked-in-place' pile.

(d) In suitable granular soils, the 'mixed-in-place' pile may be adopted. With this process, injection mortar is injected into the soil both as the auger penetrates to the required depth and during withdrawal, when the auger's rotation is reversed. Soil is not removed from the ground.

drag bit   A type of drilling *bit* consisting of short blades on a body provided with water courses that direct the drilling fluid on to the blades to keep them clean. Drilling is essentially by scraping, assisted by a jetting action of the drilling fluid against the bottom of the hole, and is best suited to clayey poorly consolidated strata.

Dräger multi-gas detector   This comprises a hand-held bellows pump of 100 cm^3 capacity with a socket into which Dräger tubes can be inserted. The tubes are made of glass into which different chemicals can be sealed which react with a wide range of gases and vapours. For measurement, the seal is broken at each end of the tube which is then inserted into the pump socket and the bellows operated a prescribed number of times depending on the gas being detected. Reaction between the gas and the particular chemical used in the tube causes a numerical value to be displayed via a colour indication which gives a measure of the gas concentration. Manufactured by Drägerwerk AG, Lübeck, Federal Republic of Germany.

drag fold   A secondary fold structure associated with a shear surface, and caused by the frictional restraint of the surfaces against one another. The direction of closure of the folds lies in the direction of motion, and so allows the sense of movement to be established.

dragline   A type of excavator comprising a power unit mounted on tracks with a bucket set with teeth located on a jib and controlled by ropes which allows the bucket to dig into material a considerable distance below or in front of the power unit and dispose of it to one side or load into transport. Suitable for excavating a wide range of materials with a wide range of bucket sizes; usually used where length of reach is important and where it would be difficult for other types of excavators to get close enough to the working surface.

drainage intensity   An estimate of the number of stream or river courses serving a particular area. The intensity of drainage is a reflection of the runoff from the surface rocks, and thus gives an estimate of their composition.

drainage path (d)   In soil-consolidation problems, the drainage path is equal to the thickness of the layer subject to *consolidation* where drainage is allowed from one side only, or equal to half the layer thickness if drainage from both sides of the layer can take place.

drainage pattern   The outline pattern of the stream and river courses in an area. This pattern is often controlled by structural features such as igneous bodies, faults, sedimentary dips, etc. It is most often employed as a line of investigation in *photogeology*.

drained test   *See drained triaxial 'smear' test; shear strength; triaxial compression machine.*

drained triaxial 'smear' test   A *drained triaxial test* wherein the normal soil specimen is replaced by two obliquely truncated right cylinders of porous stone with matching faces on to which a thin layer of *gouge* is spread. The angle of the faces is designed to be at $45° \pm \phi'/2$, where ϕ' is the assumed residual angle of internal friction. In view of the very short drainage paths involved, pore-water pressures dissipate rapidly during the shearing stage and the test can be performed in much the same time as a conventional undrained test. It is also possible to determine the drained residual

78

shear strength parameters of the gouge where only a small amount of the material can be recovered.

drained triaxial test   *See **drained triaxial 'smear' test; shear strength; triaxial compression machine.***

drawdown   Drawdown is the lowering of the groundwater level effected by pumping from a well or series of wells *(Figure D.3)*. The amount of drawdown for a given quantity of water pumped depends on the boundary conditions (whether confined (artesian) or unconfined (gravity) conditions apply), ground *permeability*, *aquifer* thickness and penetration of well into the aquifer.

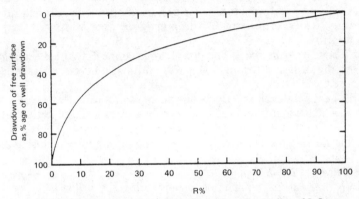

Figure D.3   Drawdown of free groundwater surface. R = Radius of Influence

dredge sampler   A marine bed sampler comprising a container which can be dragged across the sea bed to scoop up material lying at or near the surface. Usually capable of sampling any type of material, from unconsolidated deposits to hard rock.

dredgers   In general, marine equipment used for excavating and removing material from river and sea beds to provide increased water depths for vessels or creating new water channels such as canals and harbour approaches. Types include buckets or bucket-wheels, grabs, draglines, dippers, stationary suction, trailer suction, cutter suction and hopper suction dredgers.

drift   (1) Loose or uncemented materials that form a covering to harder bedrock. Geologically, these materials are usually sediments, particularly of glacial origin. The term was employed originally for glacial materials that were considered to have been deposited from floating drift ice. (2) Mineral workings that are by means of tunnels rather than vertical shafts. Hence drift mine. (3) Slow temporal changes in the strength of the geomagnetic field, or in the value recorded by any instrument whose sensor undergoes slow change. *See glacial deposits.*

drift deposit A body of loose or uncemented material, usually of glacial origin. *See drift; glacial deposits.*

drift map A geological map that records the disposition of superficial deposits rather than the underlying bedrock.

drift sheet (1) A body of *drift* that is laterally continuous over an area. Drift sheets composed of *till* are often homogeneous, but this is not implied by the use of the term. (2) When a geological map is issued in both solid and drift versions, the drift version is sometimes referred to as the drift sheet.

drill collar A heavy length of *drill pipe* placed immediately above the drill *bit* to add weight to the bit and so improve the drilling rate.

driller The man in charge of a drilling rig and crew, who operates the controls.

drill pipe, or drillstring The pipe or rods connecting the drilling *bit* to the drill and through which the drilling fluid circulates when used.

drill stem permeability test A test mainly suitable in deep *boreholes*, in which water is allowed to flow from a sealed-off test section into the empty *drill string* via a mechanically operated valve, and monitoring the pressure recovery in the section after reclosing the valve.

drillstring *See drill pipe.*

drillstring anchor A method of anchoring the *drillstring* used in *wireline drilling*, developed by Fugro of Holland, by inflating an armoured-rubber-bag packer, built into the string, against the *borehole* wall *(Figure D.4)*.

driven and cast-in-place Franki pile *See Franki piling systems.*

driven and cast-in-place pile A pile formed by driving a temporary or permanent *casing* into the ground until the required penetration or set is obtained and then filling it with plain or reinforced concrete. A variety of proprietary types are available, some of which allow an expanded base to be formed to increase the carrying capacity of the pile. *See pile foundations.*

driven pile A pile which is driven into the ground by means of a hammer acting either at the head of the pile or internally, or by a vibrator. *See pile foundations.*

driving cap A temporary cap placed on top of a pile during driving to distribute the blow over its cross-sectional area and to minimise damage to its head.

driving emf The difference between the structure–electrolyte potential and the anode–electrolyte potential in a galvanic anode system of *cathodic protection*.

drop hammer A weight, normally made of metal, used to drive piles

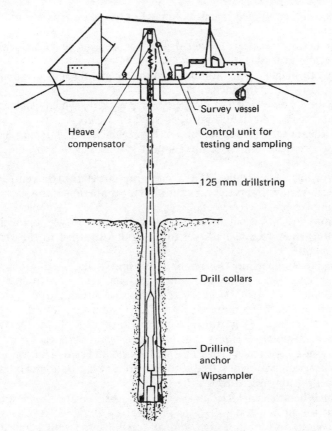

Figure D.4 Drillstring anchor

by raising with a winch and allowing it to fall under the action of gravity on to the top of the pile.

drop or stroke, piling The distance through which the *piling hammer* or ram travels.

drop sampler, or drop corer A sampler which can be dropped through the water and driven into the sea bed by weights to obtain samples of the top few metres of the sea bed material. Essentially comprises a weighted tube with a hardened metal core cutter at the base to facilitate penetration. A core catcher located above the cutting shoe aids sample retention and a one-way valve at the top of the tube prevents flushing out of the sample during retrieval. *See Cambridge sampler; Kullenberg sampler; Moore free-fall corer.*

dry density–moisture content relationship *See soil compaction.*

dry density of soil The mass of soil particles dried to constant weight, contained in unit volume. Expressed in units of Mg/m^3 or lb/ft^3.

dry hole, or duster A well drilled for oil, gas or water with an insignificant yield.

dry unit weight The ratio between the weight of solids in a material and its total volume.

ductile Possessing the property of *plasticity*, which enables the material to be continuously formed or drawn.

ductile support A ductile support is one which on failure may lead to severe deformation of the support system but not to immediate collapse.

dumpling The ground left as a temporary measure in the centre of an excavation to serve as an abutment for timbering of the surrounding faces.

dumpy level A levelling instrument used in land surveying which has the telescope and its 'level tube' rigidly attached to the vertical spindle.

Dupuit–Forchheimer assumption A simplifying assumption that water flow is purely horizontal and uniformly distributed with depth to allow *Darcy's law* for one-dimensional flow to be applicable when more intricate flow systems are solved.

Reference: 'Flow through porous media', R. J. M. De Wiest, *Fundamental principles of groundwater flow*, 1969.

duricrust A hardened crust formed in arid and semi-arid regions by the precipitation of salts due to the surface evaporation of moisture drawn upwards through the soil.

Dutch mantle cone–friction sleeve *See penetrometer (apparatus).*

Dutch sounding test A static probing test designed to assess soil conditions for pile design, and comprising a rod with a cone and a surrounding tube which is forced into the ground. The cone can be advanced independently of the tube, thus allowing end resistance and local soil friction to be measured. This mechanical arrangement is now generally superseded by an electric cone containing load cells which allow continuous recording of cone resistance and soil friction to be measured. Also developed for use with *wire line drilling* equipment for use in marine investigations. *See also cone penetration test (CPT).*

Reference: *The Penetrometer and Soil Exploration*, G. Sanglerat, Elsevier, 1972.

dye dispersion and float tracking A method used for determining the flow characteristics of coastal or river waters where this information is required for the optimum siting of intake and outfall pipes and sewers. Float tracks are used to give information about the current regime, and fluorescent dyes such as rhodamine WT or rhodamine B are injected to study the dispersion, dilution and flow directions.

dynamic analysis In essence, an extension to finite element analysis whereby a structure subjected to vibration is modelled mathematically to consider relevant details such as its foundation, subsoil, support, machinery and operating speeds. A computer analysis then determines deflections at its natural frequencies and whether the structural response is acceptable for the proposed running conditions.

dynamic consolidation A field technique, initially known as *heavy tamping*, which involves the improvement in the mechanical properties of the ground by the repeated application of very-high-intensity impacts on the ground surface. The method consists of dropping pounders weighing tens of tonnes from a crane or specially constructed machine, from heights up to about 40 m on a pattern dependent on the depth and type of material to be compacted. The French engineer Louis Menard introduced methods of estimating the required degree of compaction by in situ testing using the *Menard Pressuremeter* and monitoring the results by in situ pore pressure measurements. Very approximately, the depth of material compacted (in metres) is equal to $(Mh)^{1/2}$, where M is the pounder weight in tonnes and h is the pounder drop in metres.

dynamic point resistance The average pressure acting on the conical point in the standard *dynamic probing test*.

dynamic positioning system A system allowing a ship to remain on station without anchoring. It normally comprises four thrusters mounted fore and aft, the thrust from which is computer controlled so as to exactly balance the external forces on the ship from wind, waves and currents. It normally allows the ship to continually change its heading relative to the wind and sea conditions so as to minimise heave and pitch of the vessel, or can be set to remain on one heading.

dynamic probing test A simple *in situ* test for determining soil strength. Standardised tests include the DPA and DPB tests. They generally comprise a pointed probe attached to extension rods which are hammered into the ground with a 63.5 kg mass hammer falling 0.75 m. The number of blows per unit penetration of the probe gives a measure of soil resistance. The DPA test is performed in a borehole and casing or drilling mud is used to eliminate friction along the extension rods. the DPB test is performed without drilling mud or casing, but friction can be estimated by measuring torque required to twist the rods. Light dynamic probing equipment is frequently used in Central Europe, where methods have

83

been standardised in West Germany (German Standard DIN4094) and in Bulgaria (Bulgarian State Standard 8994-70). *See also Mackintosh boring and prospecting tool.*

dynamic resistance   The standardised result of the *dynamic probing test.*

Dywidag Gewi-pile   A drilled and grouted pile of small diameter with relatively high load-bearing capacity. It was developed from anchoring techniques, the load-bearing element being a 50 mm diameter bar.

E

Earthfirm GVS   A sodium silicate-based grout.

earthflows   Generally similar to *mudflows* but limited to the slow movement of soft weathered soils that develop typically at the toe of the slides. Earthflows are slower-moving than mudflows and are transitional between them and *flow slides.*

earth Megger tester   A portable instrument used to pass a test current between an earth *electrode* under test and a current electrode. The potential drop across the test electrode and a separate intermediate electrode is balanced by a voltage which is generated from a hand- or battery-driven AC generator via a current transformer across a digital resistor system. The out-of-balance current caused by the differences of potential is rectified and passed to a centre-zero moving coil micro-ammeter which when zeroed allows of the measurement of resistance in ohms of the electrode to earth. The instrument can also be used for shallow-depth *electrical resistivity* surveying.

earth pressure   The pressure or resultant force exerted by a mass of soil under the action of gravity or any superimposed loads. The magnitude of the pressure exerted against a retaining structure such as a retaining or basement wall is dependent for a given set of conditions on the stress history of the soil and on whether the soil is constrained against movement such that no deformation or displacements occur (the at-rest state); whether the retained soil is allowed to expand by movement of the structure away from the soil (the active state); or whether the soil is compressed by movement of the structure towards the soil (the passive state). *See also active earth pressure; Coulomb's earth pressure theory; Culmann line; K_0; K_a; K_p; passive earth pressure; Rankine's earth pressure theory.*

earth pressure balance shield   A tunnelling shield incorporating a protective steel cylinder within which a cutter wheel is sealed off by a steel bulkhead, compressed air or fluids thus being rendered unnecessary for its operation. A rotating cutter excavates the soil and continuously pushes it back into the cutter chamber by the thrust of the machine, and compresses it against the bulkhead. At the same time the soil is removed by an Archimedean-screw soil conveyor to a tank at the rear of the machine, where it is mixed to a slurry with water and pumped to the surface.

earth pressure cell A cell introduced into the ground to allow of measurement of the stress field in its vicinity. Cells comprise diaphragms and piston sensing elements, the deflection being recorded by *strain gauges* or vibrating wires. *See also* **contact stress transducer**.

earth pressure cell action factor *See action factor*.

earthquake focus The actual point of release of stored strain energy. This may be from shallow depth within the crust (a few kilometres), down to considerable depths of several hundred kilometres within the upper mantle. *See earthquakes*.

earthquakes Potentially damaging ground movements produced by sudden releases of energy along faults in the Earth's crust. Modern earthquake theory proposes the Earth's crust to consist of about a dozen discrete interlocking tectonic 'plates' (*see* **plate tectonics**) of about 100 km or so in thickness, floating on the Earth's semi-molten mantle. Where they meet, friction temporarily locks them in place, which causes stresses to build up near their edges. Eventual fracture of the rock produces sudden releases of energy, allowing the plates to resume their motion. The intensity of energy release depends on the length of time that movement of the plates is suppressed. Hence an often-quoted maxim 'the longer the wait, the bigger the bang'.

The *Richter scale of magnitude* is widely used to define the size of an earthquake, since magnitude correlates statistically with the total amount of energy released and thus is related to the potential total damage that can occur. The magnitude as measured by C. F. Richter, using a standard horizontal Wood–Anderson seismograph, is of the form $M = \log_{10} A$, where A is the trace amplitude in micrometres for an epicentral distance of 100 km.

The intensity of an earthquake is a measure of the earthquake's destructiveness at a specific location on the Earth's surface and will decrease away from the source fault owing to attenuation of the propagated seismic waves. The scale of intensity widely used is the *Modified Mercalli (MM) scale*, established in 1931 and given in *Table 18*. The standard for intensity in Japan, adopted in 1949 by the *Japan Meteorological Agency (JMA)*, is given in *Table 19*. The approximate relationship between this scale and the MM scale is $I_m = 0.5 + 1.5 I_J$, where I_m = the MM scale and I_J = the JMA scale. The *MSK scale of earthquake intensity* proposed by Drs S. V. Medvedev, W. Sponheuer and V. Karnik is based on the perception of

85

human beings and the effect on everyday surroundings, buildings and alteration of groundwater systems. The scale includes twelve categories, which are approximately similar to the MM scale. The Tangsham earthquake ($M = 7.7$) of 27 July 1967, and its principal aftershock ($M = 7.2$), which occurred 15 hours after the main event, resulted in the loss of life of over 650 000 persons in north-east China. This was the second greatest earthquake disaster in recorded history following the 1556 Shensi Province Chinese earthquake, in which at least 830 000 persons lost their lives. It is estimated that earthquakes have resulted in the loss of life of over 70 million persons during the history of the human race.

See also, earthquake focus; earthquake seismology; earthquakes . regional catalogue; earthquake zone; epicentre of an earthquake; hypocentre (or focus) of an earthquake; plate (lithospheric plate); plate tectonics.

earthquake seismology   The study of earthquakes, their resulting seismic waves and the inter-relationship with the structure of the earth. *See earthquakes.*

earthquakes—regional catalogue   A biannual catalogue of worldwide seismic events published by the International Seismological Centre in the UK. A summary is included of information concerning each event which has previously been published in the *Bulletin of the International Seismological Centre*, including epicentral estimates, origin times and dates, geographical co-ordinates, focal depths, estimates of magnitude and damage reports from various sources. *See earthquakes.*

earthquake zone   An area of the Earth's surface that is frequently affected by *earthquakes*. When considered on a global scale, the epicentres of major earthquakes are found to lie in linear belts that define the boundaries of major structural units or plates (*see plate (lithospheric plate)*). It is now known that earthquakes are generated at the margins of these plates, and only rarely elsewhere. Shallow-focus earthquakes occur at diverging plate boundaries, whereas deep foci are associated with convergence, and the passage of one plate beneath the other. In this case the earthquakes occur down an inclined zone to a depth of several hundred kilometres, and so the corresponding epicentral zone is of a similar width.

earth roads   Where necessary for temporary access works or for permanent low-cost roads, the traffic may be required to run directly on the *subgrade*, which is either left in its natural state or is graded and compacted. The surface application of some form of binder (e.g. soil, tar or bituminous emulsion) may be made to limit softening by the ingress of surface water or to reduce scour by wind. For more permanent low-cost roads, surface dressing may be applied comprising, for example, a thin bituminous spray on which stone chippings are spread and rolled in.

eccentricity of foundation loading The distance from the point of application of a force to the centre line of the base of a foundation.

echo sounder A ship-borne instrument for measuring water depth. It works by transmitting short ultrasonic sound pulses and measuring the time taken for the pulses to reach the sea-bed and be reflected back to their source.

eductor well point system, or ejector well point system A system used for dewatering excavations and similar to the *well point* system, except that the groundwater is sucked into the eductor by a vacuum produced by a nozzle and venturi system fed by a high-pressure ring main, the groundwater being pumped away through a separate header pipe to discharge. Eductors are not limited in suction lift as are well points and are best suited for deep excavations in stratified soils where close spacing is necessary.

effective angle of internal friction (ϕ') The effective angle of internal friction is a *shear strength* parameter with respect to *effective stress* in the equation

$$\text{shear strength } (\tau_f) = c' + \sigma' \tan \phi'$$

where $c' =$ the effective cohesion; and $\sigma' =$ the effective pressure.

effective cohesion (c') The effective cohesion intercept is a *shear strength* parameter with respect to *effective stress* in the equation

$$\text{shear strength } (\tau_f) = c' + \sigma' \tan \phi'$$

where $\sigma' =$ the effective pressure; and $\phi' =$ the effective angle of internal friction.

effective diameter *See particle size.*

effective grain size The effective grain size is the 10 per cent diameter (designated as the D_{10} size) on a grading curve obtained from a *sieve analysis*—i.e. the size where 10 per cent of the material is finer and 90 per cent coarser. The *permeability* of a soil has been found to be related directly to the proportion of 'fines', and an approximate value of permeability can be estimated using the *Allen–Hazen formula*.

effective overburden pressure The *in situ* effective vertical pressure existing prior to sampling or excavation.

effective porosity That part of the pore and void space in a saturated material in which movement of water takes place due to gravitational force. *See also specific yield.*

effective stress Effective stress is the difference between the applied (total) stress and the induced pressure in the pores of a material. Soil has a skeletal structure of solid material with an interconnecting system of pores which are either partly or wholly filled with water. Upon application of a load to the soil, the (total) stress is carried by both the soil skeleton and the pore water. The pore-water pressure acts with equal intensity in all directions and the stress carried by the soil skeleton alone is thus the difference between the total applied stress (σ) and the pressure set up in the pores (μ). This is termed the effective stress and is written ($\sigma - \mu$). Also, as water has no strength, the soil skeleton deforms while the pore water is displaced. This action continues until the resistance of the soil structure is in equilibrium with the external forces, the rate of dissipation of the water being dependent on the *permeability* of the soil mass and the physical drainage conditions. Professor Bishop stated the principles of effective stress in the following simple hypotheses:

(a) Volume change and deformation depend not on the total stress applied, but on the difference between total stress and the pore water pressure—i.e.

$$\text{effective stress } (\sigma') = \sigma - \mu$$

(b) Shear strength depends not on the total stress normal to the plane considered, but on the effective stress—i.e.

$$\text{shear stress at failure } (\tau_f) = c' + \sigma' \tan \phi'$$

where c' = apparent cohesion; ϕ' = angle of shearing resistance.
The concept of effective stress has been of vital importance in soil mechanics problems, in particular those where structures are stressed over long periods of time (e.g. slope stability problems). *See also **pore-pressure parameters A and B**.*

Reference: 'The principle of effective stress', A. W. Bishop, text of lecture to N.G.I. 1955, *Teknisk Ukeblad*, **106**, No. 39, pp. 859–863, 1959.

effective velocity The field velocity of groundwater movement through water-bearing strata as determined by dividing the volume of groundwater passing through a unit cross-sectional area by the *effective porosity* of the ground.

effective well radius The effective well radius for a well or well point without a filter surround is taken as half the outside diameter of the well screen, and as half the outside diameter of the filter when one is provided.

effluent streams *See losing streams.*

EFT process A method of alluvial grouting involving the driving of H-piles adjacent to each other, removing them sequentially and filling the resultant spaces by grouting under pressure.

E_h The symbol given to the *redox potential* of a solution, measured in volts relative to a standard hydrogen electrode.

ejector well point system *See eductor well point system, or ejector well point system.*

elastic analysis The calculation of stresses, strains or displacements within a body or medium on the assumption that the laws of elasticity are obeyed. In geotechnics this assumption is commonly made for foundations or excavations in stiff soils or rocks, provided that they are not too heavily fissured. In practice, although the assumptions are clearly not met in detail, the results are a useful approximation to stress distributions, etc. that are otherwise difficult to estimate.

elastic constants Coefficients of proportionality between *stress* and *strain* in an elastic material. It can be shown that only two fundamental constants are required in an ideal isotropic material (Lamé constants). In experimental practice it is more convenient to measure other constants such as the *Young's modulus*, which may be expressed in terms of the fundamental constants.

elastic heave *See heave (elastic).*

elasticity The property of a material such that deflections are linearly related to load and are recoverable. This implies that unique combinations of loads lead to unique deflections and, hence, that the measurement of deflection allows load to be calculated. Most rocks and stiff soils can be assumed to be elastic at low loads, once fissures have closed.

elastic limit The limiting stress beyond which any additional loading will cause permanent deformation of a material.

elastic material A material which has elastic properties. In practice, an upper limit to stresses will exist beyond which the material ceases to be elastic. Its further behaviour may then be either brittle or plastic, and so the terms 'elastic-brittle' and 'elastoplastic' are employed.

elastic modulus The elastic modulus of a soil can be determined from the stress–strain curve obtained during an undrained *triaxial compression test* using a cell pressure equal to overburden pressure. It is not considered to be as accurate as moduli determined from *plate bearing tests* or from back analysis of existing structures. *See also modulus of elasticity, or Young's modulus.*

elastic strain *Strain* which is proportional to *stress* and which is recovered on removal of the load.

electrical capacitance tide gauge *See automatic tide gauge.*

electrical conductivity The reciprical of *resistance.*

electrical log Any record obtained using an electrical down-hole logging sonde or surface resistivity surveying equipment.

electrical resistance tide gauge *See automatic tide gauge.*

electrical resistivity A geophysical method of surveying based on the measurement of the electrical resistance of soils and rocks. Basically it consists of passing an electrical current through the ground via two current electrodes and measuring the potential difference between two potential electrodes placed in the path of the current. From a knowledge of the measured current and voltage an *apparent resistivity* for the ground can be calculated. Various arrays for the electrodes can be used, of which the most common are the Wenner, Schlumberger, dipole and modified Schlumberger. The methods can be used to detect changes in soil or rock type and groundwater and determine the depths at which these changes take place, provided that a small number of boreholes are sunk to provide correlation for the electrical data. The most frequently used configuration is the Wenner configuration, which comprises four equally spaced electrodes in line *(Figure E.1)*. By moving the

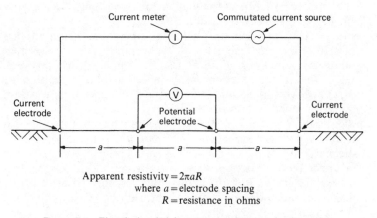

Apparent resistivity $= 2\pi aR$
where $a =$ electrode spacing
$R =$ resistance in ohms

Figure E.1   Electrical resistivity surveying. Wenner configuration

whole array forward, usually by a distance equal to one electrode separation (known as constant separation traversing), it is possible to detect lateral changes in the stratification. Another common configuration is to expand the electrode separation about a fixed central point in order to provide information on vertical changes in resistivity. With this method it is possible to assess the depths at which changes of strata occur.

Various curve-matching techniques are available for interpretation of the electrical resistivity data obtained, which are discussed in detail by H. M. Mooney and W. W. Wetzel in *The Potentials about a Point Electrode and Apparent Resistivity Curves for a Two, Three and Four Layer Earth* (University of Minnesota Press, Minneapolis, 1956) and by H. M. Mooney and E. Orellana in

Master Tables and Curves for Vertical Electrical Sounding and Layered Structures, p. 34, Table 125 (Interciencia, Madrid).

Electrical resistivity equipment can vary from the simple *earth Megger tester*, where the required information is limited to fairly shallow depths (i.e. up to about 15 m), to sophisticated equipment, where ground resistivities to depths of about 5000 m can be determined.

Electrical resistivity techniques can also be employed in boreholes and can detect subtle changes in lithology, lateral correlation of layered materials of different resistivities and the detection of permeable rocks.

See Table 10.

electrode   A conductor (metal or carbon) by means of which electric current passes to or from an *electrolyte*.
Reference: CP 1021: 1979.

electrolyte   A liquid or the liquid component in a composite material such as soil, in which electric current flows by movement of ions.
Reference: CP1021: 1979.

electromagnetic absorption   Attenuation of electromagnetic radiation due to its passage through the atmosphere or when it strikes an object. *See absorption.*

electromagnetic prospecting   A method of locating conducting bodies by using an alternating electromagnetic field to induce electric currents in them. The field is generated artificially, and the resulting currents are then detected by the fields they themselves generate. The method is commonly used in the location of ore bodies by airborne apparatus.

electromagnetic spectrum   The display of all electromagnetic radiation, which is characterised by wavelength, frequency and amplitude and moves at the velocity of light.

electronegativity   A measure of the tendency for an element to form covalent bonds.

electro-osmosis   The movement of liquid through a porous medium produced by the application of an electric potential. The applied current causes the liquid to migrate from the positive (*anode*) pole to the negative (*cathode*) pole, and this has a special application in civil engineering works for the drainage of fine-grained soils (silts and fine sands), which is difficult to achieve with conventional dewatering methods. Thus, if an electric potential is applied to a line of wells acting as cathodes and a line of rods acting as anodes, the groundwater will migrate to the wells, where it can be removed by suction. Electro-osmosis also causes a *base exchange* to take place in the soil, resulting in an increase in soil strength. This phenomenon has been used to advantage to increase the stability of earth slopes (Casagrande, 1952; Fetzer, 1967).

elevation head (1) The height above datum level. (2) In hydraulics, the height of the free surface of a fluid above datum level or the maximum height which a particle of the fluid on the free surface can fall under the action of gravity—i.e. a measure of its potential energy. *See also* **total energy of a fluid.**

elongation ratio, flatness ratio, shape factor Where a is the largest, b the intermediate and c the smallest dimension of a particle *(Figure E.2)*, then

elongation ratio $= q = b/a$
flatness ratio $= p = c/b$
shape factors $= F = p/q = ca/b^2$

Reference: 'The measurement of particle shape and its influence in engineering materials', G. Lees, *Jl Br. Granite Whinstone Fedn*, **4**, No. 2, 1964.

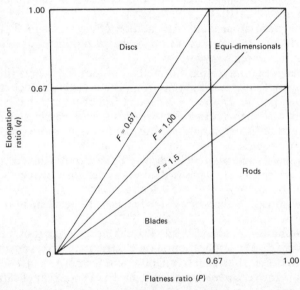

Figure E.2 Definition of shapes

elutriation A method of separating a material into its constitutent sizes by passing a liquid or gas through it, thus allowing the individual particles to settle according to their size and *Stokes's law.*

embankment A mass of compacted soil or rock placed to carry a transport system (e.g. a road or a railway line) above existing ground level or to dam the flow of liquids (an embankment dam).

Associated engineering problems include avoidance of overstressing the foundation soils, proper design for short- and long-term stability of the side slopes and adequate compaction to allow of satisfactory design of the surface pavement.

embankment dam *See embankment.*

emissivity The ratio between the amount of energy given off by a body and the amount given of by a 'black body' at the same temperature.

emulsion grouting Emulsions such as bitumen or a combination of bitumen with soap or casein or clay, in water, are used for the stabilisation of fine-grained soils. Mix-in-place methods are normally adopted for stabilising *subgrade* soils to maintain stability or load-bearing capacity, or by injection methods where watertightness is a requirement under dams or in tunnel excavations.

end-bearing pile *See bearing pile.*

engineer For the execution of a contract, the person or consulting firm, etc., appointed from time to time by the employer, and notified in writing to the contractor to act as engineer for the purposes of the contract, to ensure that the construction, completion and maintenance of the works are adequately supervised.

engineering geological map A geological map on which is recorded information of particular significance to soil or rock engineering. Typically this might be data on rock head depth and condition, weathering, jointing, etc., although the use to which such a map is to be put will largely determine the data actually recorded. Standard symbols to be used on such maps have been suggested.

Reference: 'The preparation of maps and plans in terms of engineering geology', *Qly Jl Eng. Geol.*, **5**, 293–382, 1972; BS 5930: 1981, Code of Practice for Site Investigations.

engineering geology The application of geological principles and methods to solve problems in geotechnics. In particular, engineering geology considers the effect of geological structures on large engineering works, and also the use of geological data to aid site investigations and to prepare engineering geological maps. As an academic discipline, engineering geology is distinct from both soil mechanics and rock mechanics.

engineering geomorphology The application of landform studies to problems in geotechnics. Particular attention has been paid to slope development and the historical evolution of coastal areas in the context of engineering applications.

engineer's representative Defined in the British ICE Conditions of Contract as the resident engineer or assistant of the engineer, or clerk of works appointed from time to time by the employer or the engineer and notified in writing to the contractor by the engineer to watch and supervise the construction, completion and maintenance of the *work*.

***Engineering News* formula** A design method introduced by A. Wellington-Engineering News Publications NY in 1893 for driven piles:

$$Q_\mathrm{f} = \frac{WH}{S+C}$$

where Q_f = the ultimate bearing capacity of the pile (ton); W = the weight of the driving hammer (ton); H = the drop of the hammer (in); S = the pile set per blow (in); and C = a constant representing energy losses in the driving system (in). Subsequent record has shown that the method is inherently incorrect, as the formula is based on the assumption that Q_f is a direct function of the energy delivered to the pile during the last blows of the driving process and that the energy transmission from hammer to pile and soil is instantaneous, whereas in fact the distribution of this energy with time at and after impact and the magnitude and duration of the peak impact forces are more important.

enlarged pile base An enlargement of the base area of a pile, formed (a) with a base larger than the shaft of a preformed pile; (b) *in situ*, by driving a plug of concrete into the surrounding ground; (c) *in situ*, by undercutting (under-reaming) the soil at the base of a bored pile.
   Reference: CP 2004 : 1972.

epicentre of an earthquake The point on the Earth's surface immediately above the location of the *hypocentre (or focus) of an earthquake* where the earthquake originated. *See earthquakes.*

equal area net, or Lambert net A grid used in *stereographic projection* which has the property that equal areas on the projection sphere remain equal on the grid, which corrects the foreshortening that would otherwise occur around the edges and does occur on the equal angle or *Wulff net*. Equal area nets are employed when it is desired to display information on density of distribution (for example, of joints), especially when this information is to be analysed statistically. Equal angles are not projected equally from different parts of the net, and so Lambert nets are not commonly employed for graphical analyses.

equilibrium moisture content The equilibrium moisture content is the eventual moisture content that a subgrade soil reaches in course of time after the construction of a road pavement. The moisture content of a subgrade soil under an impervious covering will tend to equilibrate with its ambient environment such that a dry soil will tend to wet-up owing to upward migration of groundwater, and a wet soil will tend to dry out if replenishment of water from, for example, rainfall is cut off. This is an important consideration when flexible road pavements are being designed from *California Bearing*

94

Ratio (CBR) tests, as they are greatly influenced by the soil moisture content.

equilibrium swelling ratio   *See swelling ratio.*

equilibrium well formulae   Well discharge formulae for steady discharge condition that assume uniform recharge at the periphery of the *cone of water table depression*. Basic formulae exist for *gravity flow* (unconfined aquifer) and *artesian flow* (confined aquifer) conditions.

Reference: *Groundwater and Wells*, Johnson Division, Universal Oil Products Co., Minnesota.

equipotential   A surface or line over or along which the value of some potential does not vary. In geotechnics the most commonly encountered equipotentials are the gravitational field and its dependents, such as fluid elevation and pressure heads. A particular use of equipotentials occurs in the flow of water through soils and rocks, since such flow will be along lines perpendicular to equipotentials, a fact that is employed in graphical flow nets.

equipotential line   *See flow net.*

equivalent continuous sound level   The average *sound* level over a given time period and the notional steady level which over a given time period would deliver the same total sound energy as the fluctuating sound levels actually existing:

$$L_{eq} = 10 \log_{10} T \int\limits_{0}^{T} \left(\frac{P_A(t)}{P_0} \right)^2 \, dt$$

where L_{eq} = the equivalent continuous sound level in dBA over a period T; $P_A(t)$ = the measured instantaneous 'A' weighted sound pressure P_A varying with time t; and P_0 = reference sound level. The period T on construction sites is normally taken to be 12 h (0700–1900) unless otherwise stated and is then written as L_{eq} 12. *See also decibel; noise; reflected sound; sound attenuation; sound level; total sound level.*

equivalent radius   The radius of that spherical particle that would have the same fall velocity in sedimentation as a given particle of some arbitrary shape.

erosion   The removal of topsoil by the action of wind or by flood water, and the weathering or undercutting of soil and rock cliffs by the action of tides, rainfall, wind and ice.

excess pore pressure   The pressure of water in a soil or rock over and above the steady state value. Excess pore pressures result from changes of load on the soil, and dissipate over time to allow of a return to the steady state. These transient flows lead to changes in

95

effective stress and so to the time-dependent volume change termed *consolidation*.

extensometers Devices for determining ground movements and basically consisting of wires or cables running in protective tubes and tensioned between anchors and a sensor head. Ground movements cause deflection of the wires, thereby changing their tension, which can be sensed by manual or automatic *transducers* in the sensor head.

F

face drain A drain constructed in the face of a slope in a pervious soil to control seepage. It may comprise a layer of coarse stone or gravel backed by a graded filter or plastics filter fabric or stone-filled mattress backed by filter material.

face shovel A type of excavator having a bucket on an arm that allows it to dig into a face in front of and above the level of the machine.

face waling, or face piece A timbering term used for a waling at the end of a trench excavation. The face waling is supported by side *walings* and together with the end *struts* supports the end face of the trench.

facing wall A liner constructed against the face of an excavation and supported by the main timbering. The liner is usually made of precast concrete units or cast-*in-situ* concrete panels which are left in place after construction, often with an asphalt tanking layer applied to provide the necessary waterproofing of the basement walls to be cast against it.

factor analysis *See multivariate.*

factor of safety The reduction factor applied to the 'failure condition' to achieve satisfactory service of a structure, foundation soil or earthwork, etc. The actual factor of safety applied will depend on the problem and the degree of certainty existing on the accuracy of the available strength parameters of the material involved and of the loadings and stress distributions.

falling-head permeability test *See pumping-in permeability test.*

falls Failure of soils in steeply cut faces in, for example, pits and trenches dug for foundations when only short-term stability is required. Stress relief (tension cracks) or shrinkage cracks occur, and failure takes place near the base of the free-standing columns of soil bounded by the crack system, causing the soil mass to fall forward into the cut *(Figure F.1)*.

fan, building A temporary screen erected on the side of a structure during construction or maintenance work to protect people and property from falling objects.

Figure F.1   Falls

fan shooting   An arrangement of seismic geophones such that they all lie on the arc of a circle whose centre is the shot point. This arrangement allows all the geophones to be equidistant from the source, and thus variations in arrival times are attributable to lateral variations in the geological structure.

fast-break   *See revert.*

fatigue   A breakdown of the original structure of a material, leading to fracture and eventual failure due to multiple strains or shocks at a much lower level of *stress* than would be required to cause failure under normal loading conditions.

fault   A fault is defined by the International Society for Rock Mechanics as a fracture zone along which there has been recognisable displacement, from a few centimetres to a few kilometres in scale. The walls are often striated and polished (slickensided) as a result of the shear displacement. Frequently rock on both sides of a fault is shattered and altered or weathered, which results in fillings such as *breccia* and *gouge*. Fault widths may vary from millimetres to hundreds of metres.

  Reference: International Society for Rock Mechanics. Commission on Standardisation of Laboratory and Field Tests, 1977.

Federal Aviation Agency (FAA) Classification   A soil classification system originally based on *mechanical analysis, plasticity* characteristics, expansive qualities and the *California Bearing Ratio (CBR) test,* for use in evaluation of soil supporting either flexible or rigid pavements in different climatic conditions. It was subsequently simplified and is now based on granulometric composition and plasticity characteristics of soil.

Fellenius solution   A method which can be used for the determination of *ultimate bearing capacity* (q_u) of purely cohesive soils. In *Figure F.2* the full-line circle represents the critical failure surface for a long surface footing on a cohesive soil, from which the expression $q_u = 5.5c$ can be deduced (c = soil cohesion).

FEMALE   A computer program acronym standing for 'Finite Element Modelling and Analysis Language for Engineers'.

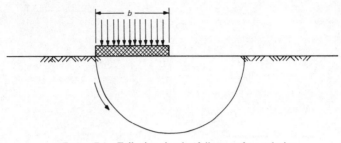

Figure F.2   Fellenius circular failure surface solution

field capacity   A term used in irrigation for the amount of water that can be retained by a soil against the force of gravity and equivalent to the term *specific retention*.

field density tests   Various methods exist to determine field or *in situ* density of naturally occurring soils or fill materials for the control of compaction in *embankments* and in connection with the design of road and airfield *pavements* and their *subgrades*. All are based on measuring the weight of a representative sample of the soil, the method of measuring the volume of which determines the type of test:

(a) The *sand replacement method*[1] consists of excavating a small smooth hole and filling it with a graded sand from a standard pouring cylinder.

(b) The *core cutter method*[1] depends on driving a cylindrical cutter into the soil without significant change of density and retaining the sample inside it so that the known internal volume of the cylinder is completely filled.

(c) The *weight in water method*[1] is applicable to soil or rock when representative samples occur in discrete lumps which will not disintegrate in water. It is restricted in practice to cohesive soils or rock specimens.

(d) The *water displacement method*[1] is an alternative to the weight in water method, with similar limitations.

(e) The *rubber balloon method*[2] is essentially a water replacement method, with a rubber membrane retaining the water.

(f) *Nuclear methods*[3] use radioactive materials. *See gamma-gamma log* and *gamma-ray density gauge*.

(g) The *water replacement method for coarse material* basically comprises excavating a hole large enough to get a representative sample, lining it with polythene sheeting and measuring the amount of water required to fill it.

Notes: 1. Specified in British Standard 1377: 1975. 2. Described in ASTM D2167. 3. Described in ASTM D2922.

field permeability tests *In situ* tests made to obtain data relating to seepage problems during construction, usually in connection with excavations into permeable strata, and seepage problems through dam foundations; and to determine the *consolidation* rate of soils in connection with the construction of *embankments* on soft soils where an increase in *shear strength* would be critical. Field permeability tests are normally made by either adding water to a *borehole* by pumping or pouring and carrying out constant-head or falling-head tests, or by removing the water by pumping or baling and carrying out constant-head or rising-head tests. Tests may be made on single boreholes, but greater accuracy is obtained by pumping from a well and recording the drawdown of the water table in observation holes sunk on lines radiating outwards from the wells. A theoretical exposition of the methods of calculating permeability for the different types of test carried out in a variety of soil profiles is given by M. Juul Hvorslev in 'Time lag and soil permeability in groundwater observations' (*Bulletin No. 36*, Waterways Experiment Station, Vicksburg, Mississippi, April 1951), and field instructions for carrying out the tests together with methods for calculation are given in *Earth Manual*, US Department of the Interior, Bureau of Reclamation (tentative edn, 1951; 2nd edn, 1974). *See also* **permeability; pumping-in permeability tests; pumping-out permeability tests.**

field vane test A test similar to the *laboratory vane test* but using equipment enabling vane tests to be made in the bottom of a *borehole* or by jacking or driving the vane to the required depth directly into the ground by use of a light boring rig or tripod.
Reference: BS 1377 : 1975, Test 18.

fill, or made ground Any material used to infill depressions, cavities, mines, etc., used in land reclamation schemes, used to raise natural ground above flood water level, or used in the construction of *embankments*. Controlled filling involves placement of fill in thin layers and compacting it with specialist plant to ensure maximum density and strength. For domestic rubbish tips, controlled tipping involves spreading the rubbish in layers and covering each layer with inert soil to prevent the spread of disease and the possibility of spontaneous combustion. *Building Research Establishment Digest 222* (February 1979) discusses problems of building on existing and new fills, such as settlement, health or environmental hazards, chemical attack and foundation design.

fillable porosity The amount of water that can be stored in an unconfined *aquifer* per unit rise in water table and per unit area. It is less than the *specific yield*, owing to hysteresis effects.

filler Filler is used in the manufacture of asphalt and consists of inert material finer than BS sieve No. 200—e.g. cement, limestone dust or *fly-ash*. Its function is to fill in the voids and change the temperature–viscosity relationship of the bitumen binder.

filling of discontinuity One of the ten parameters selected to describe discontinuities in rock masses, being material that separates the adjacent rock walls of a discontinuity and is usually weaker than the parent rock. Typical filling materials are sand, silt, clay, breccia, gouge, nylonite. The term also includes mineral coatings and healed discontinuities—e.g. quartz and calcite veins.

Reference: International Society for Rock Mechanics. Commission on Standardisation of Laboratory and Field Tests, 1977.

filter fabrics *See permeable synthetic fabric membranes.*

filters (1) Materials for removing harmful matter such as bacteria from water to make it drinkable or useful for other purposes. Filters may be mechanical or chemical in action, including, for example, sand filters, vacuum filters, porous stone and metallurgical coke, clinker or broken stone for bacteria filter beds in sewage works. (2) Protective *granular blankets* comprising one or more layers of free-draining sand–gravel material placed on a subgrade or around a well casing to allow water to seep or flow from the soil but at the same time to prevent the washing-out of fines and erosion of the soil.

A filter material should have a grading such that its voids are sufficiently fine to prevent particles of the protected soil from washing into the filter, yet at the same time be more permeable than the protected soil in order to act as a drain. It should also be such that significant segregation does not occur during placing. These criteria are normally satisfied by the following rules:

(a) *Piping* (criterion to prevent particles of the protected soil washing into or through the filter)

Rule 1 $\dfrac{\text{the 15 per cent size of the filter}}{\text{the 85 per cent size of the protected soil}} = 5$ or less

Rule 2 $\dfrac{\text{the 50 per cent size of the filter}}{\text{the 50 per cent size of the protected soil}} = 25$ or less

Rule 3 The filter should be smoothly graded; in particular, gap graded materials should be avoided.

(b) To ensure sufficient permeability

Rule 4 $\dfrac{\text{the 15 per cent size of the filter}}{\text{the 15 per cent size of the protected soil}} = 5$ or more

Rule 5 The filter should not contain more than 5 per cent of the material finer than No. 200 sieve, and fines should be cohesionless.

100

(c) To prevent segregation

Rule 6   The coefficient of uniformity of the filter.

$$\frac{\text{the 60 per cent size of the filter}}{\text{the 10 per cent size of the filter}} = 20 \text{ or less}$$

Rule 7   The maximum size of the filter material should be less than about 75 mm.

Natural variations usually result in a soil stratum or a fill material having a range of gradings. Where a filter is placed against variable material, the filter should be designed so that the more finely graded zones of the material are prevented from washing into or through the filter. Hence, in general, rules 1 and 2 should be applied to the finest gradings of the material to be protected. However, where the material to be protected contains a large percentage of gravel or larger-size particles, it is considered prudent to apply rules 1 and 2 to a finer fraction of the material. The grading curve of the fraction of the material finer than 10 mm should be used.

References: *Earth Manual*, 2nd edn, 1974, US Department of the Interior, Bureau of Reclamation; *Foundation Engineering Handbook*, H. F. Winterkorn and H. Y. Fang, Van Nostrand Reinhold, 1975.

fineness modulus   A granulometric term used widely in concrete technology. The fineness modulus is equal to 1 per cent of the sum of the cumulative percentages of material retained in a sieve analysis using US sieve Nos. $1\frac{1}{2}$, $\frac{3}{4}$, $\frac{3}{8}$, 4, 8, 16, 30, 50 and 100.

finished ground level   The level of the final surface of the ground adjacent to a building.

Reference: CP 101: 1972.

finite strain   In engineering elastic analysis, the assumption is always made that strains are infinitesimal. When it is required to analyse the deformation of geological bodies such as folded strata, it is found that the strains involved are no longer small. This has led to the theory of finite strain. Owing to the possible complexity of strain patterns in finite strain, simplifying assumptions are usually made, the most common of which is that the strain has been homogeneous or *affine*.

first arrival   The first seismic wave to arrive at a geophone. This is the wave that has travelled fastest around the ray path, and the analysis of first arrivals is the basis of the refraction method of seismic prospecting.

fish   A foreign body in a *well* or *borehole* which cannot be easily removed—often the drill *bit*. The act of removal is termed 'fishing' and in deep well drilling can involve the use of sophisticated equipment.

fissured clay Clay containing a network of hair cracks which subdivide the soil into angular fragments—e.g. London Clay.

fissures Extensive *cracks*, breaks or fractures in rock or soil.

fitting of trend surfaces *See multivariate.*

fixed groundwater Groundwater held in the pores of very fine-grained soils and rocks such that either it is permanently trapped by the high attractive interstitial forces or it can move so slowly when pumped that it is not normally available as a source of water.

flash set The premature stiffening of a cement slurry, causing it to become plastic and non-pumpable.

flat dilatometer A device developed by Professor Morchette of Italy to measure *in situ* geotechnical properties of soil. It comprises a stainless steel blade with a thin flat circular expandable steel membrane on one side which can be pushed, jacked or driven into the ground. The blade is connected to a control unit on the surface by a nylon tube containing an electric wire running through the *penetrometer* rods. The membrane is pressurised at 20 cm depth intervals, and the A pressure is read which just moves the membrane, together with the B pressure which moves its centre 1 mm into the soil. From this information is deduced the dilatometer modulus, the materials index and the horizontal stress index.

Reference: '*In situ* tests by flat dilatometer', S. Marchetti; *ASCE Jl Geotech. Eng. Div.*, March 1980, pp. 290–321.

flat jack method of uniaxial stress measurement The flat jack method consists of cutting a slot in a rock exposure, measuring the amount of closure of the rock into the slot as the stresses in the rock are relieved, and then measuring the pressure required to jack the sides of the slot apart until its initial width is restored.

Reference: 'The use of the flat jack installed in a sawcut slot, in the measurement of *in situ* stress', B. F. Wareham and B. O. Skipp, *3rd Int. Cong. Rock Mechanics*, 1974.

flatness ratio *See elongation ratio; flatness ratio; shape factor.*

flexible pavements Road or airfield *pavements* having little or no tensile strength, formed by covering stabilised soils, rock fragments or hardcore with a waterproof wearing surface of bituminous material. Prestressed concrete pavements, such as those designed by Freysinnet for the runways at Orly airport in France, are considered as being flexible. *See also pavement design.*

flexural rigidity *See pavement tester.*

flexural slip Relative movement between adjacent layers in a geological sequence when that sequence is folded in a direction perpendicular to the layering.

flexural slip folding Folding of layered rocks in which the necessary relative movement between layers is accommodated by the process of flexural slip. In such folding individual beds retain their original thicknesses and may show shear marks or *slickensides* on their boundaries.

flint   *See chert.*

floating floor slab   *See ground slab.*

floating foundation   Also referred to as a *buoyant foundation* or *compensated foundation*; the foundation of a structure for which the bulk weight of the soil removed for its construction is about equal to the total weight of the structure. Thus, the net pressure exerted to the ground by the structure—i.e. gross weight less weight of overburden removed—is very small, a condition which is often adopted where soils of low strength and/or high compressibility exist. In some cases, as, for example, in the construction of structures with basements, or dry docks, it is necessary to ensure that sufficient weight or some form of anchoring is provided to avoid *flotation* of the structure where a high groundwater table may exist.

float-operated tide gauge   *See automatic tide gauge.*

float tracking   A method of tracing water movement involving the introduction of readily indentifiable material which moves with the water, thus directly indicating its path. Standard float assemblies comprise vaned drogues constructed from canvas, wood, metal or plastics connected to a surface buoy by thin rope or wire, the length of which determines the depth at which the droge will lie and follow the water motion.

flotation   (1) A condition to avoid in constructing foundations such as basements below the water table, where the building weight must always be in excess of the maximum possible uplift (buoyancy) of the groundwater. (2) In the mining industry, a method of concentrating ore or separating ore particles from the gangue by flotation in water or other liquids having specific gravities appropriate to the ore–gangue composition.

floury   A description applied to some fine-textured soils which have a high silt fraction, the silt particles possibly containing flocculated clay of silt grain size. When dry, the material has a smooth dusty texture.

flow failure   Flow failure is defined relative to liquefaction as a form of slope movement involving the transport of earth materials in a fluid-like manner over relatively long distances, at least tens of metres.

Reference: 'Definition of terms related to liquefaction', *ASCE Jl Geotech. Eng. Div.*, **104**, Pt GT9, September 1978.

flow fold   A style of folding in which the beds show thickness changes between the noses and limbs of folds, as a result of plastic flow during deformation.

flow index (I_f)   *See Atterberg limits and soil consistency.*
flow lines   *See flow net.*
flow net   A graphical method that allows the Laplace equation for two-dimensional flow of a liquid through a porous medium,

$$\frac{\partial^2 h}{\partial x^2} + \frac{\partial^2 h}{\partial y^2} = 0$$

to be represented by patterns formed from two families of curves— i.e. *flow lines* and *equipotential lines*, intersecting at right angles *(see Figure F.3)*. The method allows of the practical solution of seepage problems which would otherwise be extremely time-consuming mathematically and in some cases insoluble. *Figure F.3* shows a simple flow net illustrating the flow path of water seeping below a sheet pile cut-off wall. The equipotential lines are lines along which

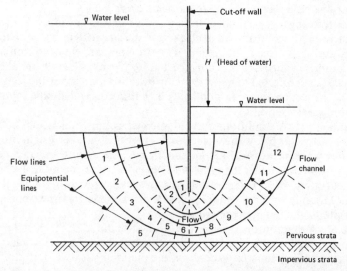

Figure F.3   Graphical method for seepage computation using a flow net

the energy level is constant—i.e. water in piezometers installed at any point along the line would rise to the same level. The flow lines give the direction taken by the seepage water. The flow net is constructed by trial and error so that the set of curves intersect at right angles and form curvelinear squares.

The volume of water seepage can be estimated from the flow net by using the expression $q = kh(n_f/n_d)$, where $k =$ the coefficient of permeability; $h =$ the total head loss; $n_f =$ the number of flow lines; $n_d =$ the number of equipotential drops (i.e. the number of spaces between the equipotential lines). The term n_f/n_d is called the *shape factor*, and in *Figure F.3* it is equal to $\frac{5}{12} = 0{\cdot}417$.

Flow nets can be constructed where the soil is anisotropic and the horizontal permeability k_h is different from the vertical permeability k_v, by constructing the flow net on a transformed section using an effective permeability equal to $(k_h k_v)^{1/2}$.

Similar principles to those given above for estimating cross-sectional flow can be used for plan flow.

flow slide A style of rapid *mass movement* characteristic of cohesionless materials. It occurs mainly in soils that have a high undrained brittleness and consequently suffer considerable loss of strength on movement. Flow slides may attain velocities of several metres per second, and can travel considerable distances. As a result they are often very destructive; for example, the flow slide that originated in a failure of colliery waste at Aberfan, Wales, in 1966.

flow till *See till.*

flow values *See tension cracks.*

fluidity The reciprocal of viscosity.

fluid potential *See total energy of a fluid.*

fluvio-glacial deposits *See glacial deposits.*

fluxgate gradiometer Portable equipment for continuous recording of lateral variations in the vertical gradient of the Earth's magnetic field. It tends to give better definition of shallow sub-surface anomalies by automatically removing the regional magnetic gradient. It is generally designed for sensitivity to features within 2–3 m of the ground surface. *See also proton magnetometer.*

fluxgate magnetometer *See magnetometers.*

fly-ash The ash residue resulting from the burning of pulverised coal in electricity power stations.

flying shore A timbering term for a *strut* or series of struts in the same vertical plane framed together to provide a horizontal restraint between two walls to prevent lateral movement of one or both walls towards each other. No part of a flying shore takes a bearing on the ground.

Reference: CP 2004: 1972.

folding wedges A timbering term for wedges used in pairs, overlapping each other and driven in opposite directions in order to hold or force apart two parallel surfaces.

Reference: CP 2004: 1972.

follower, or long dolly A piling term for a removable extension which transmits the hammer blows to the pile when the pile head is to be driven down below the leaders and out of reach of the hammer.

Reference: CP 2004: 1972.

foot block A timber pad used to spread the load from a ground prop or *side tree.*

force majeure Circumstances, such as wars, beyond the control of a contractor, pleadable as an excuse for non-completion of a contract or for over-running the allotted contract period.

formation factor In geophysics, the ratio between the *electrical resistivity* of the formation and the electrical resistivity of the pore water. This is related to *porosity, relative density* and grain shape.

formation level The finished level of an excavation prepared to receive the foundation of a structure.

FORTRAN Computer language acronym for FORmula TRANslation.

foundation That part of the structure in direct contact with and transmitting loads to the ground.

Reference: CP 101 :1972.

fracture log A well log that records the cumulative amplitude of wave arrivals produced by a sonic source during a discrete time interval. Fracture zones show up as they attenuate the acoustic energy.

Franki bored pile *See Franki piling systems.*

Franki drilled pile *See Franki piling systems.*

Franki Miga pile *See Franki piling systems.*

Franki piling systems (a) The *driven and cast-in-place Franki pile (see Figure F.4)* is formed by driving a steel *casing* to the required penetration or resistance by dropping a heavy cylindrical plunger on to a plug of aggregate or dry-mix concrete at the base of the tube. When the required depth is reached, the casing is held, the plug is expelled and semi-dry concrete is added which is forced out to form an enlarged base. The reinforcement cage is then placed in position and the shaft formed by adding further concrete, which is continually tamped as the casing is withdrawn. The method provides for good adhesion between shaft and soil and a high end bearing capacity. Pile diameters range from 350 mm to 650 mm, with working loads up to 2 MN.

Variations on the above include the *vibrated shaft Franki*, where the shaft is formed with 125 mm slump concrete and vibrated by a vibrator clamped to the top of the casing; the *composite Franki*, where the shaft can be either of precast concrete or of cast-in-place concrete but with a permanent casing; and the *special Franki*, which incorporates preboring techniques.

(b) The *Franki bored pile:* With this system a *borehole* is formed with a coring tool operating inside temporary lining tubes. On

106

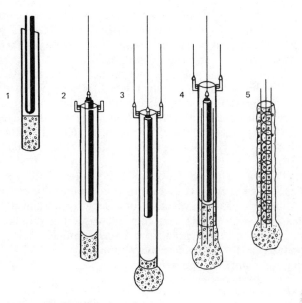

1 Consolidating about 1 m of
aggregate to form a solid plug
2 Driving the tube
3 Forming the base
4 Forming the shaft
5 Completed pile

Figure F.4   The Franki driven-*in-situ* pile

completion of boring, semi-dry concrete is added which is expelled
to form an enlarged base in a similar way to that for the driven pile.
The subsequent placing of the reinforcement cage and formulation
of the shaft are carried out in a similar way to that for the driven
pile.

(c) Large-diameter *augered piles* (450–2100 mm diameter) with
under-reamed bases (2250–5400 mm diameter) can be constructed
to carry working loads up to 20 000 kN.

(d) The *Franki Miga pile* system is principally for use in
underpinning works and comprises precast concrete sections
measuring 300–350 mm in width and 700 mm or more in length
which are jacked into the ground, with the structure to be
underpinned being used as the necessary reaction. Each unit has a
steel-lined hole about 50 mm in diameter running vertically
through its centre, with its first section provided with a pointed end
piece to aid penetration. As each unit is added, lengths of steel are
inserted in the central hole and grouted in position to make an
effective joint between neighbouring sections.

107

(e) **Barrettes** are large oblong piles formed under bentonite using diaphragm walling techniques, ranging in size from 2200 × 600 mm to 2200 × 1800 mm and carrying working loads up to 20 000 kN. They can be formed in any type of ground.

(f) The *Franki drilled pile* is a pile formed by boring with a continuous hollow-stem flight auger which is screwed into the ground by the hydraulic rotary head of the machine, the concrete being subsequently pumped down the hollow stem as the auger is withdrawn from the ground. Piles of 450 mm and 600 mm diameter are available to carry working loads to 800 kN and 1200 kN, respectively.

free-end triaxial test   The use of lubricated end platens in all types of *triaxial compression tests*. It gives improved uniformity of *stress* and deformation and allows of reduction in length of specimen (to a height/diameter ratio of 1) and thus increases the number of specimens that can be prepared from a single sample.

free-fluid index (FFI)   *See nuclear-magnetism log.*

free-fluid log   *See nuclear-magnetism log.*

free water   *See groundwater.*

frequency of discontinuities   A parameter used in the description of rock core recovered from a *borehole*, and defined as the number of natural discontinuities intersecting at 1 m length of recovered core.

Reference: International Society for Rock Mechanics. Commission on Standardisation of Laboratory and Field Tests, 1977.

friction circle   A technique used in the analysis of slope stability and earth-pressure problems to give the direction of any vector representing an intergranular pressure of obliquity ϕ to a circular rupture arc *(see Figure F.5)*. The radius of the 'friction circle' or 'ϕ circle' is made equal to $R \sin \phi$, which is concentric with the circular rupture arc of radius R.

front-end loader   A type of excavator comprising a purpose-built shovel mounted on track or pneumatic-tyred operating power units. It is often used in situations such as in quarries, where material excavated at the face requires to be transported a short distance to a lorry or conveyor.

frost action   A general term for damage caused by freezing and thawing of moisture in materials and in structures of which they are a part or with which they are in contact.

frost boil   *See frost heave.*

frost front   The position in the ground at which freezing takes place at any particular time. It is usually approximately parallel to the ground surface or other surface from which heat is being extracted.

frost heave   The raising of a surface owing to the formation of ice in the underlying soil. Also known as *frost boil.*

frost susceptibility test   A test carried out on road pavement subsoil

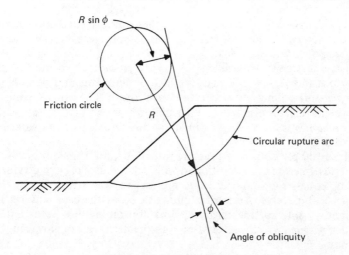

$R \sin \phi$

Friction circle

R

Circular rupture arc

ϕ

Angle of obliquity

Figure F.5 Friction circle

material to determine its susceptibility to heaving when frozen. The test described in *Road Research Laboratory Report LR 90* requires compaction of the sample into a 100 mm diameter by 150 mm high cylindrical mould. After extrusion the sample is placed in an ambient temperature of $-17°$ C while the underside is kept in contact with water at $+4°$ C. The heave of the specimen is recorded daily over a period of 250 hours, the total heave giving the following order of susceptibility:

Heave (mm)	Frost susceptibility classification
0–12.7	Non-susceptible
12.8–17.8	Marginally susceptible
>17.8	Frost-susceptible

frost-susceptible soil Materials in which sufficient ice segregation will occur to result in *frost heave* or heaving pressures when the requisite moisture and freezing conditions exist. Frost-susceptible materials generally include silts, silty fine sands, chalk, limestone, burnt colliery shale and pulverised fuel ash where 40 per cent or more of the particles pass a 75 μm sieve. *See frost susceptibility test.*

frozen-ground phenomena A general term for geomorphological processes and features that result from freeze–thaw processes.

Typical frozen-ground phenomena include sorted nets and stripes, *solifluction*, ice lensing (*see ice lenses*) and wedging (*see wedge failure*), and involutions.

full-face diamond bit drilling   A *rotary probe drilling* method in which the *bit* has a full cutting face of diamonds and permits of *open-hole* drilling and integration with core drilling with a medium-weight rig. It is also used for forming regular holes for *in situ rock mechanics* testing—i.e. *in situ* stress measurements. Air, water or mud flush may be used.

full, or 100 per cent, pore-pressure ratio   A ratio defined in relation to *liquefaction* as a condition in which the change of pore pressure, Δu, equals the initial effective minor principal stress, σ_{3c}, or the vertical effective overburden pressure, σ_v, as the case may be. In many contexts this term replaces the ambiguous term *initial liquefaction*, which was used by some geotechnical engineers in the past.

Reference: 'Definitions of terms related to liquefaction', *ASCE Jl Geotech. Eng. Div.*, **104**, Pt GT9, September 1978.

fully softened strength of clay   The lower or 'critical state' condition determined by measuring the strength of remoulded *normally consolidated* clay.

Fulmer tension meter   A portable mechanical gauge for measuring tension in cables and ropes. The gauge is clamped to the cable or rope and a horizontal force is applied to it, the tension being determined by bending of the frame, which is relative to the applied force.

Fundex pile   A pile formed by screwing into the ground a steel drive tube that is simultaneously being pushed down by hydraulic rams. When the required formation level has been reached, concrete and reinforcement are placed in the tube, which is then withdrawn. *See also pile foundations*.
Reference: CP 1021: 1979.

G

gabions   Stone-filled rectangular wire-mesh boxes which may be laid like bricks to form self-standing structures (e.g. *retaining walls*) or laid as mattresses to form flexible linings for coastal and river bank protection. Either form may be grouted with bituminous sand mastic to either fully penetrate the stone or just grout the surface, depending on the degree of impermeability required.

gaining stream   Streams (also called *influent streams*) that have their flows increased by submerged springs or seeps at bed level.

gal   A unit of acceleration equal to 10^{-2} m/s^2. The nominal gravity of the Earth is 980 gal.

gallery A drainage tunnel constructed from the face of a slope or from a shaft excavated adjacent to the slope. It may provide access to transverse adits, inclined bored drains or grouting points, and is constructed at a gradient to allow of gravity drainage towards the portal.

galvanic action A spontaneous electrolytic cell reaction in which the metallic *anode* corrodes.
Reference: CP 1021: 1979.

galvanic anode An *electrode* used to cathodically protect a structure by *galvanic action*.

galvanometer An instrument for measuring small currents, wherein a coil suspended in a magnetic field rotates through an angle proportional to the current flowing through the coil.

gamma A unit of magnetic-field intensity, equal to 10^{-5} oersteds.

gamma-gamma log, or density log A well log obtained by measuring the back-scattered *gamma rays* from the formation produced by using a logging tool or sonde containing a gamma-ray source such as caesium-137. The secondary radiation produced is proportional to the electron density of this formation, which is itself roughly proportional to the *bulk density*. *See well logging.*

gamma-ray density gauge The gamma-ray density gauge works on the principle that the back-scattered gamma radiation from a gamma-radiation source is a function of the *bulk density* of the material surrounding the source, a detector sensitive to gamma radiation being used to measure the back-scattered radiation. A common gamma-radiation source is caesium-137, the radiation being detected by a scintillation counter such as a Geiger–Müller tube.

gamma-ray log A well log that records a plot of natural gamma-ray count. Since minerals containing radioactive isotopes tend to concentrate in clay-type soils and rocks, the method is suitable for recording the location of clays and shaly rocks. It is often used instead of the SP log in cased holes. *See also gamma–gamma log.*

gamma rays Electromagnetic radiation emerging as photons, similar to X-rays but of shorter wavelength, emitted by the nuclei of radioactive atoms during their decay.

gap grading An aggregate with both fine and coarse material but without intermediate size fractions.

gas guns *See continuous seismic reflection profiling.*

gatch A local name for material encountered in the Middle East (especially in Kuwait) consisting of lightly consolidated marine sand formed mainly from precipitated calcium and magnesium carbonate. It is chalky white in appearance, often saline and gypsiferous, and frequently used as a road-base material.

gelifluction A synonym for *solifluction*, except that it makes explicit reference to the need for frozen ground conditions, whereas solifluction can strictly refer to soil flow under non-frozen conditions.

gel time   The time taken from start to completion of gelation. It may be started by adding a *catalyst* to the main solution.

geodimeter   A surveying instrument for the precise measurement of distance. It works on the principle of transmitting a series of wave-shaped pulses of visible light to a reflector. The time taken for the return journey of the 'pulses' is measured indirectly by the method of 'phase comparison' and is capable of determining a time interval of 1 part in 30×10^{12} s, thus providing extremely accurate measurement of distance.

geodrain   A band-shaped drain strip comprising a filter paper wrapped around a plastic core which can be rapidly inserted into the ground by machine (similar to *alidrain*).

geological map   A map on which is recorded the disposition of the rocks over an area of the Earth's surface, together with information on their attitude, lithology, presumed age, etc. The significant feature of all geological maps is that, provided that the strata are assumed to be continuous, the subsurface structure may be deduced from the outcrop pattern by the application of simple geometric rules. In the United Kingdom the publication of national geological maps at various scales is undertaken by the Institute of Geological Sciences. These maps are published as either solid, drift or solid and drift sheets. The scale of primary mapping is normally 1 : 10 000 (formerly six inches to one mile) and publication is at this scale and smaller.

geology   That branch of science concerned with the constituents of the Earth's crust, their arrangement and structure, and the natural forces which tend to modify them.

Geonics non-contacting resistivity meter (EM31)   A portable instrument permitting of measurement of terrain conductivity (or resistivity) without probes, using an inductive electromagnetic technique. Measurements to a depth of about 6 m can be made as fast as the operator can walk and the technique is applicable in engineering geophysics for, e.g. detection of near-surface solution cavities in chalk and limestone.

Geonor field inspection vane tester   Portable equipment marketed by Geonor of Norway, used for rapid and simple determination of the undrained *shear strength* of clay, comprising exchangeable vanes on hand-held rods.

Geonor settlement probe   A *Borros point* type of *settlement* reference probe.

geophone   An instrument used in *seismic methods of surveying* which transforms the seismic energy at a point into an electrical voltage.

112

Normally of the moving-coil type, comprising a coil suspended in a magnetic field, the coil moving relative to the magnetic field when subjected to a shock, thus generating an electric current.

geophysical surveying Techniques generally involving the determination of variations in the physical properties of soils and rocks which, when correlated with limited *borehole* information at specific locations, can provide rapid and economic coverage of a site. Methods include measurement of electrical conductivity (resistivity); density (gravimetric); magnetic susceptibility (magnetic); and sonic velocity (seismic). *See electrical resistivity; gravity surveying methods; magnetic surveying method; seismic methods of surveying.*

Reference: BS 5930: 1981, Code of Practice for Site Investigations.

geophysics The physical study of the Earth using methods such as seismic reflection and refraction, gravity, magnetic, electrical and radiation techniques.

Geoseal grout A water-soluble resin grout produced by the Borden Chemical Co (UK) Ltd, to combat a variety of groundwater conditions.

geotechnical processes A general term for the processes used in ground improvement, such as *grouting, vibro-flotation (vibro-compaction)*, etc.

geotechnics A term currently employed to cover the fields of *soil mechanics, rock mechanics* and *engineering geology.*

geotextiles *See permeable synthetic fabric membranes.*

geothermometer A device for determining the temperature of deep-seated geothermal reservoirs by measuring the chemical composition of the hot water as it emerges from springs or wells. It is based on the solubility of certain minerals varying with temperature. The source temperature inferred by SiO_2 concentration (silica geothermometer) and/or sodium, potassium and calcium concentration (Na–K–Ca geothermometer).

ghanats Well-and-tunnel groundwater supply systems hand dug by ghanat diggers or 'moghannies'. Originating in Iran possibly around 2000 BC, they have been found in other parts of Asia, and also in Africa, Europe and South America. Also known as ganats, kanats, karviz and foggaras, the systems comprise the construction of one or more wells about 1 m in diameter and typically 40–50 m deep, from the base of which a tunnel is excavated at a slight incline until daylighting at the lower part of a valley or basin. Shafts to ground level are built at intervals for ventilation.

ghost, seismic A secondary arrival at a *geophone* that results from a reflection at the ground or water surface and thence a journey around the seismic path. On a continuous record, ghosts appear as horizons that mimic the shape of the true horizons; hence the name.

giga A prefix denoting one thousand million (10^9).

gilsonite A pure variety of *asphalt* found in North America.

girdle A distribution pattern displayed by poles on a stereogram, in which the poles are distributed randomly in azimuth, but have a consistent dip. The resulting projection is a continuous belt or girdle that is centred on the great circle of mean dip. *See stereographic projection.*

GKN driven pile The GKN driven pile is formed by driving a heavy temporary lining tube fitted with a detachable shoe, to the required penetration or set *(Figure G.1)*. The reinforcement cage is then

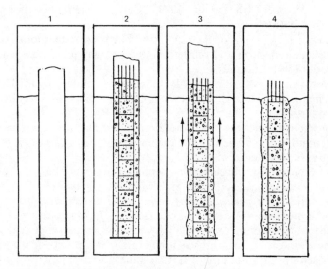

1 Temporary casing driven to required depth and set
2 Reinforcement placed and casing filled with concrete
3 Casing being vibrated during extraction
4 Completed pile

Figure G.1 Stages in construction of GKN driven pile

placed and the tube filled with concrete to above cut-off level. The lining tube is vibrated on extraction, which causes the concrete to flow easily and completely fill the space left by extraction of the tube. Pile diameters of 375 mm, 425 mm and 475 mm in lengths up to 25 m are available to carry working loads of 500–1200 kN, depending on soil conditions.

glacial burst, or joküllhaupt A phenomenon encountered on the Icelandic ice cap Vatnajoküll, which is seated on an active volcanic area. Geothermal heat causes the production of meltwater which is discharged in periodic floods or bursts. The periodicity results from the need to attain a critical head of water before the ice can be floated off its bed. The discharges involved are substantial, and the need to accommodate them has been a major constraint on the transportation system of southern Iceland.

glacial deposits Glacial deposits include all the products of glaciation and are classified according to whether material is stratified or not. *'Boulder clay'* is often used to describe unstratified glacial deposits, even though it may contain neither boulders nor clay. Thus, the term is misleading and the Scottish term 'till' is preferred. Stratified glacial deposits are mostly water-transported and may be generally classified as *'fluvio-glacial deposits'*. *See also* **drift; till.**

glacio-isostacy The concept that the isostatic balance of the continents has been changed by the growth and decay of major continental ice sheets, particularly during the Pleistocene period. The term is often used in the context of the slow uplift that follows the rapid removal of the ice load, since this rebound has resulted in the formation of many features of geomorphological interest — notably, raised shore lines and marine deposits. *See* **isostacy.**

glaciotectonic structures Deformational structures observed both in bedrock and in glacigenic sediments, which are considered to be the result of stresses generated by the flow of glacier ice over the material. Typical glaciotectonic structures include large, flat-lying overfolds which close in the direction of ice movement, major thrusting of wedges of transported bedrock, and local deformation at the boundaries of subglacial or over-ridden sediment units. Current thinking favours a mechanism which relies on a reduction of *shear strength* due to the generation of high subglacial pore pressures.

Glötzl hydraulic cell An earth pressure cell comprising a hydraulic sensing pad which is embedded in the soil, the soil stress producing pressure on this pad which is then recorded by mechanical methods.

Glötzl piezometer A type of pneumatic piezometer.
Reference: 'The Gepatsch rockfill dam in the Kauner Valley', H. Lauffer and W. Schober, *Trans. 8th Int. Congr. Large Dams*, Edinburgh, Vol. 3, pp. 636–660, 1964.

glyben An artificial frictionless clay consisting of a mixture of bentonite powder and glycerine.

Goble pile-driving analyser An electronic instrument developed by Pile Dynamics Inc., USA and marketed by Balkan Piling Ltd, UK, for measurement of pile bearing capacity during installation of the pile.

gold-198 (^{198}Au) A radioisotope of gold having a *half-life* of 64.8 h used in the form of gold chloride for radio tracing. *See radiotracers*.

Goodman jack An NX *borehole* probe, marketed by Sinco of Seattle, with movable rigid bearing plates for the measurement of rock load–deformation characteristics. Two models are available, providing bearing plate pressures of 64 MN/m^2 (12-piston) and 38 MN/m^2 (3-piston).

gouge Gouge is finely ground material within a faulted zone or along a slip surface in soil or rock. It is often highly plastic, and although it usually occurs as a very thin division along slip surfaces (often 1 mm or less), it is often possible, if sufficient care is taken, to scrape away enough of the gouge to allow *drained triaxial smear tests* for the determination of *residual shear strength* parameters of the material to be made. *See residual and peak shear strength*.

grabs Specially designed excavators or general-purpose excavators fitted with a grabbing tool to allow of mucking out of spoil from within sheeted excavations or for use in dredging operations. They may be fitted to machines having hydraulically powered arms or jibs and the grabs themselves may be power-operated.

grab sampler A sampler for obtaining samples from the sea-bed and basically having jaws which close on contact with the sea-bed (by some mechanism such as weights, springs, lever arms or cords) and engulf the bed material. Examples are the *Shipek sediment sampler* and the *Van Veen grab*.

granular blankets *See filters*.

gravel The fraction of soil composed of particles between the sizes of 60 mm and 2 mm. The gravel fraction may be subdivided as follows:

Gravel grading	Size
Coarse	63–20 mm
Medium	20–6.3 mm
Fine	6.3–2.0 mm

See also particle size.

Reference: BS 1377 : 1975.

gravel pack A graded stone filter placed in a water well between the formation and the *well screen* to increase the *specific capacity of a well*, minimise loss of fines from the formation, aid construction of the well and enable a larger screen size to be used, thus minimising incrustation. In general, either a uniform grading or a graded-grain size may be used. The former aids well development by allowing some fines to pass from the formation through the pack, thereby increasing the *permeability* of the formation adjacent to the pack; however, it is usually difficult to obtain uniform material from local sources and it is normally necessary to make up a pack material on site from available material. The ratio between the mean grain size of the gravel pack and that of the formation is termed the 'gravel-pack ratio', which is normally equal to

$$\frac{50 \text{ per cent size of the gravel pack}}{50 \text{ per cent size of the formation}}$$

Wells with a gravel-pack ratio of 4–5 generally have a high efficiency. Higher ratios are usually accompanied by lower well efficiencies, wells having ratios greater than 10 producing excessive sand.

Reference: *Water Well Technology*, M. D. Campbell and J. H. Lehr, McGraw-Hill, 1973.

gravimeter An instrument for measuring the acceleration due to the attractive force that the Earth exerts on any body minus the centrifugal force experienced by that body due to the Earth's rotation. Three basic types exist: the *beam-type gravimeter*, wherein a beam is supported against gravitational attraction by a spring; the *spring-type gravimeter*, wherein a weight is suspended on a spring and the gravitational change is determined from the restoring forces required to compensate the spring extension; and the *vibrating-string-type gravimeter*, wherein gravimetric change is determined by measuring the change in resonant frequency of a short vertical bar which occurs under the gravitational attraction. Sea-bed gravimeters are essentially remotely controlled sensitive land gravimeters.

gravitational gliding The large-scale movement of a sheet of rock during mountain building. It is supposed that the sheet moves as a coherent layer, possibly assisted by internal fluid pressure, along a major plane of transport (*plane of décollement*). Buckling of the sheet during transport produces nappes, or large-scale overfolds, and over-riding at the end of the travel produces tectonic overthrusts.

gravitational water *See groundwater.*

gravity flow *See equilibrium well formulae.*

gravity platform Gravity platforms for offshore gas/oil production units or drilling decks, and similar uses, are structures designed to rest on the sea-bed and to rely on their own weight to maintain position and resist the environmental forces of the sea and the wind. They are generally regarded as having considerable advantages over the open-jacket type of platform where they are to be situated in rough open-water conditions, as they require little or no time-consuming piling in the installation phase. *See open-jacket platforms.*

gravity surveying methods Gravity surveying methods depend on measuring very small variations in gravity at successive stations which are magnified by mechanical or optical means to obtain readable quantities. The use of gravity meters, which, although sensitive to variations in the Earth's gravity of only about 1 part in 10^8 (about 0.01 mgal, or 10^{-7} m/s^2), are restricted to locating major geological structures and large subsurface voids. *See gravimeters.*

grid roller A machine used for compaction of rock fill, coarse-grained granular fills and cohesive soil dry of its optimum moisture content, having two cylindrical rollers on which rectilinear patterns of welded steel bars are provided to allow high contact pressures to be applied to the ground.

grit A coarse sand having more or less angular-shaped grains.

gritty Having sufficient coarse angular-shaped grains of sand to give a characteristic rough feel to otherwise fine-textured soils.

gross loading intensity The intensity of vertical loading at the base of a foundation due to all loads above that level. *See also loading intensity, or unit pressure; net loading intensity.*

ground anchor A structural member that transmits an applied tensile force to competent ground, the *shear strength* of which is used to resist the tensile force. Anchors may comprise *tension piles*, *rock-bolts*, *deadmen* or high-strength steel *tendons* formed of bars, strands or wires grouted into predrilled holes, and may be either dead or prestressed by initially tensioning the anchor to the structure or to, say, a ground slab which at the same time tests the integrity of the anchor. Anchor types include cylinders filled with grout *(Figure G.2(a))*, for use mainly in rocks; cylinders enlarged by grout injected under high controlled pressure or by forcing pea gravel into the side of the anchor hole, for use in cohesive and non-cohesive soils *(Figure G.2(b))*; and cylinders mechanically enlarged at one or more positions along the length of the hole, for use in clay strata *(Figure G.2(c))*.

References: *Design and Construction of Ground Anchors*, T. H. Hanna, CIRIA Report 65, 1980; 'Rock anchors—state of the art', G. S. Littlejohn and D. A. Bruce, *Ground Engineering*, July 1975, September 1975, November 1975, May 1976.

ground beam A beam in a substructure transmitting a load to a pile, pad or other foundation.

Reference: CP 2004 : 1972.

ground failure A permanent differential ground movement capable of damaging or seriously endangering a structure.

ground freezing A technique used to stabilise fine-grained soils or coarser soils where normal methods of groundwater control might

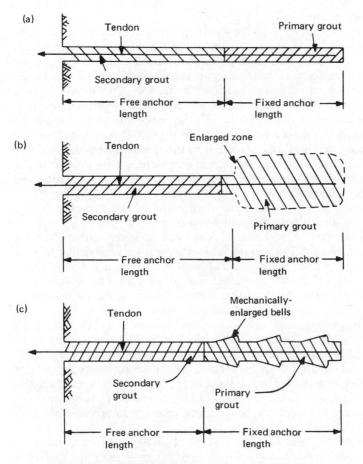

(a) *Type 1 anchor:* Cylinder filled with grout
(b) *Type 2 anchor:* Cylinder enlarged by grout injected under high but controlled pressure. Enlarged zone formed by grout permeation into granular soil. Little or no enlargement in rock or cohesive soil
(c) *Type 3 anchor:* A cylinder mechanically enlarged at one or more positions along its length

Figure G.2   Illustration of main anchor groups available in UK

cause stability problems, to allow of excavation below the water table. Freezing imparts strength and impermeability to the ground and can be carried out either by circulating cooled brine through tubes (freeze probes) in the ground or by applying a refrigerant (e.g. liquid nitrogen) directly into the probe and allowing it to evaporate at the primary point where cooling is required.

ground probing radar Equipment employing the technique of an impulse radar system for studying subsurface features such as soil and rock conditions, pipe-work and cavities. The technique is sometimes known as *electromagnetic subsurface profiling* and can be considered as the electromagnetic equivalent of the single-trace acoustic profiling method used for marine sub-bottom profiling. *See also GSSI impulse radar system; terrain conductivity meter.*

ground slab A concrete floor slab that rests on the ground within the external walls of a building (*see* CP 101: 1972). A *floating floor slab* is one that is supported directly by the ground, whereas a *suspended floor slab* is one that is designed to span across the ground and be supported by *sleeper walls* and/or the main foundations supporting the superstructure.

ground truth Data observed or collected on the ground to resolve anomalies observed by *remote sensing.*

groundwater '*Free water*' or '*gravitational water*' within the saturated zone below ground level at a pressure greater than atmospheric, which can be collected in an excavation (e.g. well or pit) or emerges naturally at ground level as a *spring*. The surface of the groundwater, or saturation zone, is known as the *water table*, or standing water level, which may vary, depending on rainfall, climatic conditions and season, etc. The zone above the water table is known as the *vadose* zone, which includes a zone known as the *capillary fringe* immediately above the water table where groundwater is drawn up by surface tension to fill or partly fill the interstices in the ground. Perched groundwater or *perched aquifers* comprise groundwater in a saturated zone which is separated from the main body of groundwater by a relatively impermeable layer of soil or rock.

groundwater cascade The descent of *groundwater* on a steep *hydraulic gradient* to a lower and flatter water table below a groundwater barrier such as a dam.

groundwater decrement Water removed from the subsurface *groundwater* body by all causes, including evaporation, *transpiration*, seepage, well pumping and stream flow.

groundwater divide The boundary of the *cone of water table depression* formed by pumping, separating the *area of influence* from the area outside.

groundwater hydrology That part of *hydrology* concerned with the occurrence, distribution and movement of water below ground level.

groundwater increment Water added to the subsurface *groundwater* body from all sources.

groundwater, perched *Groundwater* overlying a relatively impermeable soil or rock layer and separated from the main groundwater body. Its surface is termed a *perched water table.*

groundwater ridge   A ridge-shaped body of *groundwater* produced by an *influent stream.*

groundwater table   *See groundwater.*

groundwater trench   A trench-shaped depression of the *water table* produced by seepage into a stream or drainage ditch.

groundwater, unconfined   Water in an aquifer having a *water table* at equilibrium with atmospheric pressure.

grout curtain   *See curtain grouting; grouting.*

grouting   The injection of materials, under pressure, into the ground or man-made structures to increase their strength and/or reduce their *permeability.* Grouts may be categorised as chemical, suspension or emulsion systems, and include cement, sand-cement, clay-cement, slag-cement, resins, gypsum-cement, clays, asphalts, bitumens, pulverised fuel ash (PFA), and various colloidal and low-viscosity chemicals.

GSSI impulse radar system   Equipment manufactured by Geophysical Surveys System Inc, of New Hampshire, USA, for use as a shallow subsurface exploration tool for engineering applications. *See also ground probing radar; terrain conductivity meter.*

guide frame   A timber frame erected above ground level to act as a guide for runners or trench sheeting, and as a staging from which they may be driven.
Reference: BS 6031:1981.

guide runner   A runner driven ahead to form a guide for driving intermediate runners.
Reference: BS 6031:1981.

gulls   *See cambering.*

gumbo   A very sticky, highly plastic clay.

Gunite   A fluid mixture of cement, sand and water which can be sprayed through a nozzle on to a surface, such as a rock mass, to seal and protect it. Termed *Shotcrete* in Europe.

Guttman process   A modification of the *Joosten process* of grouting, comprising the addition of an alkali (sodium carbonate) to the silicate solution in order to reduce the viscosity and enable finer-grained soils to be treated.

gyp   A rock-like scale deposit in a well, normally formed by a precipitation of calcium sulphate or calcium carbonate from the groundwater.

H

half-life ($T_{1/2}$)   The time taken by a radioactive material for its activity to decay to half its original value.

hand auger Originally known as a post-hole auger; simply a boring tool which can be turned by hand to form a hole for the emplacement of a post—e.g. a telegraph pole. It is useful for site investigations, especially for light reconnaissance surveys to investigate softish cohesive soils in which boreholes can stand unsupported without *casing* and are free of stones coarser than about medium gravel-size. Boreholes are normally of 150 mm and 200 mm diameter, and limited to about 4–5 m depth, from which disturbed and 40 mm diameter undisturbed samples can be taken for identification and soil strength determination.

hardness The hardness of a material is defined in one of two ways: (1) The indentation hardness is the force required to cause indentation by a cone, divided by a constant that measures the cone shape and area. The resulting value has units of stress, and for a plastic material is similar to the plastic yield stress. (2) The so-called mineralogic hardness is found by comparison with an arbitrary reference scale, and is based on the ability of one material to scratch another. *See Moh scale of hardness.*

hardpan Part of the *'B' horizon* of the topsoil which becomes thoroughly cemented to an indurated rock-like material owing to an accumulation of calcareous or ferruginous cementing material. It does not soften when wet and it limits the downward penetration of roots and water.

hardrive pile *See West's shell piling system.*

hardware Equipment such as *computers* and *remote sensing* instruments, used for data collection and handling.

Harrison hardrock corer A small sea-bed drill containing its own battery power supply, used to obtain short undisturbed samples of core from exposed areas of bedrock.

Hazen's law *See permeability.*

head boards Boards supporting the ground at the roof of a *heading*.

head deposit A *drift deposit* produced by *solifluction* under periglacial conditions and characterised by locally derived poorly sorted, angular materials. They grade into river deposits and are thus often difficult to determine precisely.

head, elevation Head due to the potential energy of a fluid. *See total energy of a fluid.*

heading An excavation made by tunnelling.

head, pressure or static Head due to static energy of a fluid. *See total energy of a fluid.*

head, total The sum of the elevation head (due to the potential energy), the velocity head (due to the kinetic energy) and the pressure or static head (due to the static pressure) of a fluid. *See total energy of a fluid.*

headtree A horizontal member placed immediately on the head of a *dead shore*.
Reference: CP 2004 : 1972.

head, velocity Head due to the kinetic energy of a fluid. *See total energy of a fluid.*

head wave A *refraction wave* or *Mintrop* wave, characterised by entering and leaving 'high-velocity' media at the *critical angle*.

heat of hydration The heat evolved during the setting and hardening of concrete.

heave compensator A device used in marine drilling to compensate for heaving of the ship's drilling platform due to wave action.

heave, elastic The lifting of the floor of an excavation due to the elastic expansion of the ground as the weight of overburden is removed.

heaving pressures The stresses acting against a structure that result from ice formation in frost-susceptible soil, or wetting-up of a soil, or elastic expansion of a soil as the overburden pressure is removed. *See heave, elastic.*

heavy minerals Minerals of high *specific gravity* (greater than 2.9) which may be separated from the *sand* fraction of a soil by floating off the lighter grains in a heavy liquid or a range of such liquids. The importance of heavy minerals lies in the fact that they are often specific to igneous provinces and, being resistant to weathering, they can survive in sedimentary environments, thus providing evidence of the provenance of the detritus.

heavy soils Soils of firm consistency with a high clay fraction and a dense structure.

heavy tamping *See dynamic consolidation.*

held water Water held in the soil pores by capillary or adsorption forces. *See capillarity.*

Hercules piling system The Hercules piling system incorporates precast reinforced hexagonal units in various lengths joined together with a special interlocking bayonet-type joint. The bottom unit is fitted with a hardened steel cylinder 150 mm long × 60 mm diameter, having a concave face and sharp edges to facilitate penetration of sloping strata and boulders. Two standard sizes for safe bearing loads of 700 kN and 1200 kN are available. *See also pile foundations.*

HI-FIX A compact, lightweight and highly mobile offshore survey electronic position-fixing system based on the *Decca Navigator System*. The equipment comprises a master and two slave stations which can be displayed in either a Two-range or Hyberbolic form by placing the master station ashore or on the survey vessel *(see Figure H.1)*. The Two-range version gives maximum accuracy but

123

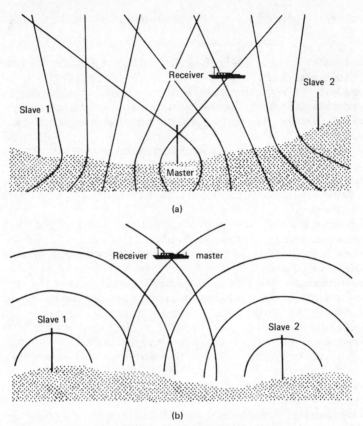

Figure H.1 Hi-fixes: (a) Hyperbolic, (b) Two-range

is usable on only one vessel at a time. With the Hyperbolic layout, a high-accuracy service is provided to an unlimited number of users simultaneously with the two slave stations and the master unit aligned along the shore and with a receiver placed aboard each vessel. The equipment may be supplemented by shipborne displays to give either simple meter readings for reference to prepared *lattice charts* or pictorial displays showing the vessel's movement on a chart. The HI-FIX Duplex system in addition provides two independent systems with continuous automatic lane identification. *See also MINI-FIX; SEA-FIX.*

high oblique (aerial) photograph   An aerial photograph taken from such an angle that it is possible to see the line of the horizon. *See also low oblique (aerial) photograph.*

Hiley formula   The Hiley formula relates energy of hammer blow to resistance of ground when piles are being driven:

$$R_u = \frac{wh\eta}{s + \frac{1}{2}c}$$

where R_u = the ultimate driving resistance; w = the weight of the hammer (ton); h = the drop of the hammer (in); η = the efficiency, depending on the ratio between P/w and the coefficient of restitution (e); s = the set or penetration of the pile per blow; P = the weight of the pile and any driving cap, dolly or anvil block; and c = the total temporary compression ($c_c + c_p + c_q$), where c_c = the temporary compression of the dolly, c_p = the temporary compression of the pile and c_q = the temporary compression or quake of the ground. The formula is a derivative of the **Engineering News formula** and suffers from the same drawbacks in accuracy as given for that method.

hinge line   (1) In tectonic usage, a line that divides an area of relative uplift from one of relative downwarping. Hence, an area of no relative movement. (2) Of a fold: the line about which the two limbs have been folded.

Hochstrasser–Weise   A large-diameter grab-type bored piling rig that utilises a continuous semi-rotary motion of the *casing* to keep it sinking as the *borehole* is advanced.

Hoek triaxial cell   An instrument that can be placed over a cylindrical core of rock and allows a uniform hydraulic pressure to be applied while subjecting the specimen to an axial compressive load. It is used for determining the strength and elastic properties of rock.

hoggin   A naturally occurring deposit of sand and gravel with just sufficient clay to bind the material together without significantly affecting its frictional properties.

homeomorphy   The concept that two or more bodies may have different origins or histories, but may exhibit similar final forms.

homoaxial folding   Folding within a given fold train or system in which the axes of the folds are subparallel.

homogeneous deformation   *Affine* deformation.

homologous forms   Forms (for example, geomorphological land forms) that obey the concept of *homeomorphy*.

horizontal movement gauges   Devices for measuring the horizontal movement of the ground and structures. Types include telescoping tubes and tensioned wire devices.

Reference: *Foundation Instrumentation*, T. H. Hanna, Trans Tech Publications, 1973.

horsepower   (1) A measure of the rate at which work is performed. The Imperial unit of horsepower (h.p.) is equal to 33 000 ft lb of work done in 1 min. 1 h.p. = 0.7457 kW. (2) *Indicated horsepower* is the theoretical power output of an engine. (3) *Brake horsepower* is the actual power output delivered by an engine and is equal to the indicated horsepower multiplied by the mechanical efficiency of the engine. (4) *Unit boiler horsepower* represents the amount of work done in converting 34.5 lb of water per hour into steam at a pressure of 1 atmosphere and at a temperature of 212° F.

hot mix recycling A process in which *reclaimed asphalt pavement (RAP)* materials, *reclaimed aggregate material (RAM)*, or both, are combined with new asphalt, and/or new recycling agents, and/or new aggregate, as necessary, in a central plant to produce hot-mix paving mixtures. The finished product meets all standard material specifications and construction requirements for the type of mixture being produced. *See cold mix recycling; surface recycling*.

Reference: *Civil Engineering*, January 1981.

hot-wire anemometer *See anemometers*.

H-pile A steel wide-flange column or other section, often rolled with a uniform section in web and flange. *See steel piles*.

Reference: CP 2004 : 1972.

humus A necessary constituent of agricultural *topsoil* comprising partly decomposed vegetable matter.

Hunter and Schuster method of slope stability analysis This method assumes a potential circular arc failure surface in a purely cohesive, normally consolidated, unfissured clay with *cohesion* varying linearly with depth. A *water table* ratio

$$M = \left(\frac{h}{H}\right)\left(\frac{\gamma_w}{\gamma'}\right)$$

is considered, where h = the depth from the top of the slope to the water table during *consolidation*; H = the height of the cut; γ_w = the density of water; γ' = the submerged density of the soil; and *capillarity* above the water table is assumed to give a fully saturated soil to ground level. The factor of safety is given as

$$F = \left(\frac{c}{p'}\right)\left(\frac{\gamma'}{\gamma}\right)N_s$$

where c = cohesion; p' = the effective vertical stress; γ' = the submerged density of the soil; γ = the bulk density of the soil; and N_s = the stability factor, obtained from charts giving N_s for values of water table ratio M against slope angle β.

Reference: 'Chart solutions for analysis of earth slopes', J. H. Hunter and R. L. Schuster, *Highway Research Record*, No. 345, 77–89, 1971.

Hush piling system Basically a drop hammer running on a leader frame mounted on a crawler crane, with the leaders, hammer and pile enclosed in a box constructed from acoustically dead material.

hydraulic conductivity A hydrologic term for *coefficient of permeability, or hydraulic conductivity*, and defined as the unit volume of water flowing through a unit cross-section of ground in unit time under a *hydraulic gradient* equal to 1. The product of the hydraulic conductivity and the aquifer thickness is called the *transmissibility*.

hydraulic fracturing (1) A method of measuring in situ stresses in *normally consolidated* clay whereby a length of *borehole* is sealed and water pumped in at incremental pressures until a sudden increase in water flow takes place, at which stage tensile failure of the soil is assumed to have occurred. (2) A method of *well stimulation* obtained by injecting a liquid under high pressure into a well in order to fracture the formation and thereby increase its *permeability*. See *hydrofracturing*.

hydraulic gradient The loss of *hydraulic head* per unit length in the direction of flow.

hydraulic head The sum of the pressure head and the geometrical height above a given reference level.

hydraulic piezometer See *continuous piezometric logging; piezometer*.

Hydrodist A marine survey instrument for distance measurement, having principles of operation similar to those of the *Tellurometer*. See also *Autotape*.

hydrofracturing A technique employed to increase the rate of flow of water or other fluid into a well. Water is pumped into the well at high pressure, and the tensile stresses so induced in the walls of the well cause the surrounding rock to fracture. With careful design, interconnecting patterns of fissures may be obtained between adjacent wells, this enabling fluid to be circulated between them. The technique has applications in the recovery of oil and the extraction of geothermal heat.

hydrogen ion concentration The number (H) of parts per 1000 of hydrogen ions in a solution and a measure of the solution's acidity—i.e. its *pH value*, where $pH = \log_{10} 1/H$. Neutral solutions have a pH of 7, while values higher or lower than 7 indicate acid or alkaline solutions, respectively.

hydrograph A graphical representation of the change in flow of water or elevation of water level with time.

hydrology The study of the physical and chemical properties of water on the Earth, including its precipitation, its movement in the ground and its eventual return to the atmosphere by evaporation.

hydrometer A graduated float, the level of which, when it is placed in a liquid, is dependent on the latter's *specific gravity*.

hydrophilic A term descriptive of that property of a substance that allows of absorption and adsorption of water.

127

hydrophobic   A term descriptive of that property of a substance that causes it to repel water.

hydrostatic head   The height of a column of stationary water that is equivalent to some given pressure.

hydrostatic stress   At any depth in a body of water, the stress will be hydrostatic—i.e. will be equal in all directions. This may be formally stated in terms of *principal stresses*, in that all the principal stresses are equal. Thus, in any medium a state of hydrostatic stress is said to exist if the principal stresses are all equal. If one principal axis is vertical, hydrostatic stress implies that the horizontal and vertical stresses are equal. In general, any state of stress may be decomposed into a hydrostatic component whose magnitude is the average value of the principal stresses, and a deviatoric component that is the extra stress in a given direction. In the case of a soil, one may further note that changes in spherical stress result in volume changes, whereas changes in deviatoric stress result in shear distortion. *See deviator stress.*

hypocentre, or focus, of an earthquake   The location at which an earthquake originates. The point immediately above the hypocentre on the Earth's surface is called the *epicentre*. *Earthquakes* may be classified as:

extremely shallow—for depth of focus < 30 km
shallow—for a depth of focus of 30–100 km
deep—for a depth of focus > 100 km

See epicentre.

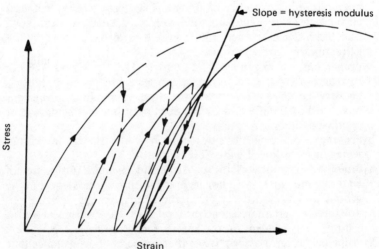

Figure H.2   Hysteresis modulus

hysteresis modulus of elasticity   The average slope of the final stress–strain curve obtained when a material is subjected to compression at the stage where the hysteresis loops from several loading and unloading stages indicate minimal further *strain* for a particular applied *stress* (*Figure H.2*).

I

[131]**I**   *See iodine-131,* [131]*I.*

ice, volumetric relation to water   100 ml of water at $0°$ C equals 109.036 ml of ice at $0°$ C—i.e. a 9.036 per cent increase in volume.

ice lenses   Ice formations in soil occurring essentially parallel to one another, generally normal to the direction of heat loss and commonly in repeated layers.

ice segregation   The growth of ice lenses, layers, veins and masses in soils, commonly, but not always, orientated normal to the direction of heat loss.

ice-side till   *See till.*

image-motion compensation   A method of film processing wherein, in order to minimise image blurring of films taken from an aircraft, the film is run backwards through the camera during exposure, to compensate for the forward motion of the aircraft.

immediate settlement   *See settlement.*

impact crushing value test   A test used to determine the crushing value to be used in the classification of chalk in relation to its behaviour as a freshly placed fill material.
Reference: *The Classification of Chalk for Use as a Fill Material*, H. C. Ingolby and H. W. Parsons, UK TRRL Report No. 806, December 1977.

impeller anemometer   *See anemometers.*

impressed current   The current supplied by a rectifier or other direct current source to a protected structure in order to attain the necessary potential for *cathodic protection.*

impression blocks   Impression blocks are often used to obtain an impression of the top of a tool lost down a borehole, before fishing operations are attempted. They are of many forms and designs. *See fish.*

inclinometer   An instrument used to monitor lateral movements of earth slopes, bulkheads, piles during driving, etc. The simplest type comprises a flexible tube placed in a *borehole*, the deflection of which is indicated by resistance to a steel rod or 'torpedo' lowered down it.

incompetent bed   If a layered composite of two dissimilar materials is buckled, it is often found that one set of layers retain their thickness, while the other set flow to accommodate changes in the geometry of the composite. This effect is observed in layered rocks, in which some beds show considerable thickness changes when traced around folds, these changes being due to pressure gradients that existed during folding. Such beds are termed *incompetent*, and the less yielding beds are termed *competent*. Typically, clays and soft sandstones are incompetent, whereas hard sandstones, limestones and lavas are competent. The term is used in a relative sense, and so does not indicate an absolute property of a particular rock type. *See competent bed.*

incremental loading plate bearing test   A test carried out to determine the *ultimate bearing capacity* of a soil by applying increasing increments of load until failure of the soil has occurred or until a specified multiple of the proposed working load has been reached. Cycles of loading and unloading can give an indication of the order of elastic deformation, and the undrained *modulus of deformation* may be determined if the rate of loading is sufficiently rapid. *See also* **constant rate of penetration test (CRP test)**.

incrustation   The clogging up of a well screen or the voids in a *gravel pack* or surrounding water-bearing formation due to a collection of material formed from precipitation of minerals normally carried in solution in the groundwater, such as calcium and magnesium carbonates, fine clay, silt or sand particles, and the presence of iron bacteria or slime-forming bacteria in the *groundwater*.

index properties   *See Atterberg limits and soil consistency.*

index property tests   *See Atterberg limits and soil consistency.*

indicated horsepower   *See horsepower.*

induced polarisation   A resistivity surveying technique whereby an electrical potential is applied across a rock and abruptly discontinued. In some rocks the potential decay is gradual rather than instantaneous *(see Figure 1.1)*, analogous to the discharge of a capacitor, the effect being known as 'induced polarisation' or the IP effect. The rate of decay can provide information on the presence of minerals in the void space, thereby allowing of the detection of ores in mineral exploration or of changes in salinity and *permeability* in hydrogeological surveys.

induction log   A well log that records measurements of *electrical conductivity* and *resistivity* produced by inducing a high-frequency alternating current of constant intensity to flow in the formation.

indurated   A sedimentary rock is said to be indurated if the pore space is filled with secondary mineral cement, so producing a strong, compact rock.

infiltration capacity, or potential infiltration rate   The rate at which water can enter soil when the supply is unlimited—i.e. when the soil

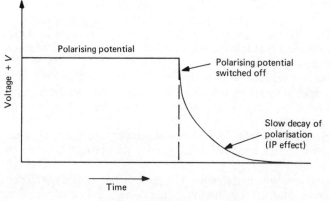

Figure I.1 Induced polarisation

is covered by a body of surface water such as a stream, lake or ponded rainwater.

influent seepage The downward movement of water from ground level to the *water table* in the *zone of aeration, or unsaturated zone*, due to gravity.

influent streams *See gaining stream.*

infra-red That part of the *electromagnetic spectrum* with wavelengths between 10^{-3} m and those of visible light.

initial liquefaction *See liquefaction.*

initial tangent modulus of elasticity The slope of the initial straight line portion of the stress–strain curve obtained when a material is subjected to *compression. See also secant modulus of elasticity.*

injection wells Wells used for the recharge of aquifers in order to limit or reverse the decline of *groundwater* levels (*see artificial recharge*).

inside clearance ratio (of a soil sampler) A ratio equal to $(D-d)/d$, where $d=$ the inner diameter of the cutting nose and $D=$ the inner diameter of the container.

***in situ* stress** At any point at depth in a body of rock there exists a state of stress due to the weight of overlying material, and possibly also as a result of residual stresses after previous deformation. This is the *in situ* stress in the rock body. The orientation and shape of the stress ellipsoid may vary considerably, being approximately spherical in stable areas, but being elongate with an inclined or horizontal major axis in areas of current mountain building. Near to surface the shape of the ellipsoid is determined by the surface topography.

instantaneous elastic strain *See creep.*

intensity of earthquake A measure of the destructiveness of an earthquake at a specific location on the Earth's surface. The

131

intensity decreases away from the *epicentre of an earthquake* owing to the attenuation of the seismic waves. *See earthquakes; Japan Meterological Agency (JMA) scale of earthquake intensity; Mercalli (modified) scale of earthquake intensity; MSK scale of earthquake intensity.*

interceptor drains   Subsoil drains used to collect and divert excess ground or surface water away from soil banks and cuttings or road pavements, which would otherwise have their stability or performance impaired.

interformational fold   A fold that involves more than one geological stratum, thus demonstrating that it postdates the strata that it deforms.

intermediate principal strain   The intermediate strain corresponding to one of the three orthogonal directions, the shear strains of which are equal to zero. *See also average stress, or octahedral normal stress; intermediate principal stress (σ_2); major principal strain; major principal stress (σ_1); minor principal strain; minor principal stress (σ_3); octahedral shear stress.*

intermediate principal stress (σ_2)   The intermediate *stress* acting on one of the three orthogonal planes where shear stresses are equal to zero. *See also average stress, or octahedral normal stress; intermediate principal strain; major principal strain; major principal stress (σ_1); minor principal strain; minor principal stress (σ_3); octahedral shear stress.*

intraformational fold   A fold that is entirely contained within one stratum, thus demonstrating that it is penecontemporaneous with the formation of that stratum.

inverted echo sounder tide gauge   *See automatic tide gauge.*

iodine-131, ^{131}I   A radioactive isotope of iodine, having a half-life of 8.05 days, that emits low-energy *gamma rays* and is used as a *radiotracer*.

ion   Either of the products which appear at the respective poles when a substance is subjected to electrolysis, whence any of the electrically charged particles which are released by dissociation in an *electrolyte*.

isochrone   A line joining points of equal excess hydrostatic pressure (pore-water pressure) within a soil during the process of *consolidation*.

isohyetal map   A map showing the rainfall distribution of an area, on which contours of equal precipitation (isohyetal lines) are drawn.

isopachytes   Lines joining points on the ground having equal thicknesses of *overburden* or equal thicknesses of a particular stratum.

isopiestic line   A contour joining points of equal pressure at the surface of a confined *aquifer*.

132

isostacy The theory of isostacy holds that at some depth (the depth of compensation) the *lithostatic pressure* is equal over the entire globe. This implies that masses of light rock 'float' higher than masses of dense rock. In particular, it is believed that elevated areas of crust such as mountain chains may be balanced by roots of similar dimensions. A consequence of the theory is that if a rock mass is disturbed, it will return to equilibrium. Such movement — due, for example to unloading by the retreat of the Pleistocene ice sheets—has left distinct geological features such as raised shore lines and other marine features and sediments. *See isostatic compensation.*

isostatic anomaly A geophysical measure of the extent to which a rock mass is out of isostatic equilibrium. Isostatic anomalies are calculated in a similar way to gravity anomalies, and are also measured in milligals. A negative anomaly implies a deficiency of mass below that required for compensation, and *vice versa.*

isostatic compensation The concept that there is some depth of compensation at which the pressure of the overlying rocks is uniform. This carries the corollary that high mountains are 'balanced' by roots that enable lighter crustal rocks to project into the lower layers. *See isostacy.*

isotopes Atoms of the same element which differ in their mass number owing to the different numbers of neutrons in their nuclei.

isotropy The state wherein all significant physical properties are equal in all directions.

isotropic (1) Having physical properties (e.g. strength, permeability) which are equal in all directions. (2) Capable of transmitting light with equal velocity in all directions.

ISSMFE International Society of Soil Mechanics and Foundation Engineering.

J

jacked pile A pile, usually in short sections, which is forced into place by jacking it against a reaction such as the weight of the structure.

jack method A method of measurement of in situ stress in rock. An area of rock is relieved of stress, and its deformation is noted. Pressure is then applied by a system of jacks to restore the rock to its original dimensions, and the pressure so required is a measure of the *in situ* stress that was relieved. *See flat jack method of uniaxial stress measurement.*

Jacob method An approximation of the Theis non-equilibrium formulae (*see Theis method*) which avoids the need to match the plotted drawdown with 'type curves'. The aquifer constants are given by:

$$\text{coefficient of transmissibility } (T) = \frac{0.1832Q}{\delta z}$$

$$\text{coefficient of storage } (S) = \frac{2.25t_0 T}{\gamma}$$

where $\delta z =$ the change in drawdown per log cycle of time; $\gamma =$ the distance from the well to the observation hole; and $t_0 =$ projected intercept of time to zero drawdown.

Reference: 'A generalised graphical method for evaluating formation constants and summarising well field history', H. H. Cooper, Jr and C. E. Jacob, *Trans. Am. Geophys. Union*, **27**, 1946.

Janbu method of slope stability analysis This method is based on the method of slices. It assumes a cylindrical potential failure surface and can accommodate a layered soil profile, the effects of tension cracks and surcharge, and the influence of partial submergence and drawdown conditions.

Reference: *Stability Analysis of Slopes with Dimensionless Parameters*, N. Janbu, Harvard Soil Mech. Series, No. 46, 1954.

Japan Meteorological Agency (JMA) scale of earthquake intensity A scale of earthquake intensity adopted as a standard in Japan and given in *Table 19. See earthquakes.*

JCL Computer language acronym for Job Control Language— used to instruct a computer how to handle specific jobs.

jetting, of water wells A *well stimulation* process involving high-velocity washing of screen openings in the *borehole casing*. This agitates the sand and gravel particles surrounding the screen and washes fines out of the formation through the screen and into the well, thereby increasing *permeability* and well efficiency.

joint A surface of discontinuity within a mass of rock. Joints may be either curved or planar, and have surface textures which can vary from polished to rough on the scale of the rock grains. The rock on either side of a joint has undergone little relative displacement, the term *fault* being used if movement has taken place. The detailed origin of joints is obscure in many cases, but they are often related to principal stress directions and so may well originate as brittle fractures in the stress fields associated with particular geological events. In sedimentary rocks many small discontinuous joints sometimes are found; such features are usually called *fissures*. A joint is defined by the International Society for Rock Mechanics as a break of geological origin in the continuity of a body of rock along which there has been no visible displacement. A group of parallel joints is called a set, and joint sets intersect to form a joint system. Joints can be open, filled or healed. Joints frequently form parallel to bedding planes, foliation and cleavage, and may be termed bedding joints, foliation joints and cleavage joints, respectively.

Reference: International Society for Rock Mechanics. Commission on Standardisation of Laboratory and Field Tests, 1977.

joint sets (system, family) Groups of joints all having a common orientation, and being similarly spaced from one another.

joküllhaupt *See glacier burst.*

Joosten process A ground injection process introduced by H. J. Joosten in 1925 for increasing the strength of soils. It involves the successive injection of sodium silicate and a strong electrolytic salt solution such as calcium chloride, which react together to form a gel and permanent solidification of the mass. The Joosten II process includes an alkali solution to reduce viscosity and facilitate penetration of finer-grained soils, while the Joosten III process provides for the injection of sodium silicate, an ammoniacal colloid and a heavy metal salt to give an even less viscous grout. *See also Guttman process; Monodur process.*

juvenile water New water of magmatic, volcanic or cosmic origin.

K

K_a The coefficient of *active earth pressure.*

K_0 The coefficient of lateral earth pressure at rest (i.e. when the soil is prevented from expanding or compressing laterally). For *normally consolidated* soils $K_0 \simeq 1 - \sin \phi'$, where ϕ' is the angle of internal friction in terms of *effective stress*, derived from Jaky's (1944) theoretical expression

$$K_0 = \frac{(1 + \frac{2}{3}\sin \phi')(1 - \sin \phi')}{(1 + \sin \phi')}$$

For overconsolidated clays this expression is no longer valid and K_0 can increase to unity for an *overconsolidation ratio (OCR)* of about 5, and up to $K_0 = 2.5$ for heavily overconsolidated soils and even higher for machine-compacted clays. The accurate measurement of K_0 is very sensitive to changes in the environmental stresses which occur when samples are taken for laboratory testing or when measuring devices are inserted into the ground. Attempts to minimise such disturbance for *in situ* testing include the development of the *Camkometer* by Wroth and Hughes (1973) and a pressure meter by Jaguelin *et al.*, 1972.

K_p The coefficient of *passive earth pressure.*

kentledge Material used to add temporary loading to a structure—e.g. to the top of caissons to assist in sinking or as dead weight in a loading test.
Reference: CP 2004 : 1972.

kicking piece A length of timber spiked to a *waling* to take the thrust from the end of a strut which is not at right-angles to the waling.
Reference: BS 6031 : 1981.

kinetic energy *See total energy of a fluid.*

king piles Piles (normally of H-section) driven along the line of a wide trench before excavation and to the full anticipated depth. They are located at *strut* intervals, and as excavation proceeds serve as supports and abutments for the struts. In difficult ground the piles may be placed and partly driven in prebored holes.

kips An abbreviation denoting 'thousand pounds per square inch' (properly kpsi). More recently just used to indicate thousand pounds, as in, for example, $2 \text{ kips/ft}^2 = 2000 \text{ lb/ft}^2$.

Kullenberg sampler A *drop sampler (corer)* used for obtaining samples from the sea-bed.

L

laboratory soil mixer Any power-driven mixer used for pulverising soil for soil testing or for blending mixtures of soil or admixtures of stabilisers and water to soil. Usually with a capacity of up to about 5 kgf.

laboratory vane test A test carried out on soft clays to determine shear strength by inserting a small cruciform vane into a sample and measuring the torque required to shear the soil when rotated. An alternative to the undrained triaxial compression test where preparation of the sample may have an adverse effect on the measured strength of the soil.

lacing A system of ties connecting one setting of *walings* or *struts* to another. This allows the weight of the lower frame to be carried as excavation proceeds.
Reference: CP 2004 : 1972.

LAMBDA A low-ambiguity Decca positioning system involving a mobile master station transmitting on two frequencies (100–200 kHz) which are received by two small fixed stations to control phase-locked transmitters.

lambda (λ) method of pile design A method for determining the frictional capacity of pipe piles driven in cohesive soils. This is given by $Q_s = \lambda(\bar{\sigma}_m + 2c_m)A_s$. The frictional capacity, Q_s, of a pile in clay is assumed to be dependent on the effective vertical stress and the undrained shear strength. $\bar{\sigma}_m$ = the mean effective vertical stress between the sea-floor and the pile tip; c_m = the mean undrained cohesive shear strength along the pile length; A_s = the surface area of the pile; and λ = a dimensionless coefficient (a function of pile penetration) ranging from 0.4 to 0.2 for piles penetrating less than 15 m, decreasing to about 0.1 for piles embedded about 60 m. *See also **alpha (α) method of pile design** and **beta (β) method of pile design**.*

Reference: 'A new way to predict the capacity of piles in clay', John A. Focht, Jr and V. N. Vijayvergiya, *Proc. 4th Annual Offshore Tech. Conf.*, Vol. 2, pp. 865–874, 1972.

Lambert net   *See **equal area net, or Lambert net**.*

Lamé constant (λ) *See **stress–strain relationships**.*

laminar flow   The motion of a fluid wherein the particles move substantially along parallel paths, as opposed to turbulent flow. Laminar flow is also termed *straight line, streamline* or *viscous flow*. The flow is always laminar when the velocity is below the critical velocity and it can be laminar when the velocity is between the lower and higher critical velocities.

land facet   *See **land system**.*

land form element   A small part or feature in a *land system*. An element is a repeated unit which shows recurrent features that distinguish it from other elements. The features taken as diagnostic depend on the nature of the land system and the use to which it is being put. *See **terrain analysis**.*

landslides   *See **classification of landslides**.*

land system   A terrain classification comprising a pattern of land forms evolved during reasonably uniform climatic and geological conditions and consisting of similar topographical units known as *land facets*. These are units of landscape, fairly distinct from the surrounding terrain, on which soils and materials are usually fairly uniform.

References: *Working Group Report on Land Classification and Data Storage*, A. B. A. Brink *et al.* MEXE Report 940, 1969; 'Application of terrain evaluation to road engineering', P. J. Beaven and C. J. Lawrence, *Proc. Conf. Road Engineering in Asia and Australia*, Kuala Lumpur, 1973.

Langelier index, or saturation index   An indicator for determining the incrustation or scale-forming parameters involved in calcium carbonate ($CaCO_3$) deposition on well casings and screens, taking into account the alkalinity of the groundwater, the concentration of ions present, the *pH value*, the temperature of the groundwater and the total amount of dissolved solids.

Reference: 'The analytical control of anti-corrosion water treatment', W. F. Langelier, *AWWA Jl*, **28**, 1936.

large-diameter bored pile   A pile, generally 600 mm or more in

diameter, formed by boring with power-driven or mechanical drilling and grabbing equipment. *See also bored and cast-in-situ concrete piles.*
Reference: CP 2004 : 1972.

laser An acronym for Light Amplification by Stimulated Emission of Radiation. A laser comprises a special high-intensity light source in a relatively small solid angle. It is used in the civil engineering industry for directional control during construction of sewers, tunnels, trenches, etc.

laser–Doppler anemometer *See anemometers.*

lateral spread Lateral spread is defined in relation to *liquefaction* as distributed extensional movement in a fractured mass in which extension of a rock or soil results from liquefaction or plastic flow of subjacent materials.
Reference: 'Definition of terms related to liquefaction', *ASCE Jl Geotech. Eng. Div.*, **104**, Pt GT9, September 1978.

laterite A red ferruginous porous clay derived from the weathering of basaltic and other lavas. It is composed mainly of alumina and iron oxide, and covers large areas of some tropical countries.

lattice charts *See circle (lattice) charts.*

LCN method of aircraft pavement design The load classification number (LCN) method of aircraft evaluation and pavement design, based on load, tyre pressure and contact area of undercarriage, pavement type (rigid, composite or flexible), aircraft category, subgrade characteristics, and allowable flexural concrete stress.
Reference: *Design and Evaluation of Aircraft Pavements*, Directorate of Civ. Eng. Dev., Department of the Environment, London, 1971.

leaching The process whereby soluble salts are removed from soil by the percolation of water.

leaders, for piling Leaders are generally constructed from steel channels or tubes and are used for locating and guiding the hammer and pile during driving, either forming part of a pile frame, or suspended from the jib of a mobile crane or supported on a mobile base or standing on the ground secured by guy-ropes.
Reference: CP 2004 : 1972.

least squares, method of A mathematical technique whereby a graph is fitted to a set of data points which do not lie exactly on it, whether as a result of error or other cause. The criterion of best fit is that the sum of the squares of the deviations of the points from the line be a minimum; hence the name. Mathematically, this is equivalent to finding a perpendicular projection of the observations vector into a vector space of lower dimension. It should be noted that other criteria of best fit are available, which lead to other types of approximation.

Leeman or CSIR triaxial strain cell An instrument for measuring the in situ triaxial state of stress, comprising a strain cell containing nine electrical *strain gauges* set in three groups at 120° to one another, which are expanded against the walls of a borehole. The stress relief is then measured as the borehole containing the cell is over-cored *(Figure L.1)*.

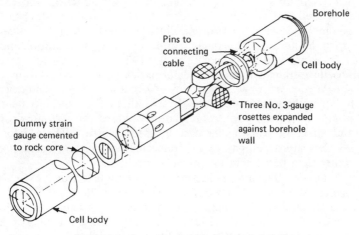

Figure L.1 Leeman triaxial cell — exploded view

level A survey instrument for determining the height of a point above a datum. It consists basically of a telescope, which can be rotated in azimuth, which is fixed in the plane of the equipotential surface by levelling an associated spirit bubble. In operation, the level of the required location is obtained by reference to a previously known height by sighting a graduated staff located at each position through the level's telescope and measuring the difference in level.

light soils Soils having a medium to coarse texture, low silt and clay content and loose *relative density*.

lignochrome A grout material manufactured from lignosulphide, a residual product resulting from the production of cellulose. Soluble lignosulphides when oxidised by bichromate result in the precipitation of a heavy metal salt, and mixtures of these chemicals form firm gelatinous masses with setting times of up to about 10 h, depending on the percentage of bichromate present.

lime column method of soil stabilisation A method developed in Sweden whereby powdered unslaked lime is mixed *in situ* with soft clay or silt with an egg-beater-shaped auger, the lime being forced into the soil by compressed air just below the horizontal blade of the auger. It has been used to increase bearing capacity and reduce settlement in a variety of situations—e.g. instead of bearing piles for light structures and sheet piles in deep excavations.

lime rock A soft clayey limestone occurring in Florida and Georgia, USA.

limit analysis methods of slope stability Methods adopting relatively new techniques using the concept of a yield criterion and the associated flow rule that considers the stress–strain relationship. The methods are discussed and illustrated by examples in *Foundation Engineering Handbook*, edited by H. F. Winterkorn and H. Y. Fang, Van Nostrand Reinhold, 1975.

limited flow strain, or limited flow deformation Defined in relation to *liquefaction* as flow strains or deformations that commence following liquefaction but are arrested after a finite displacement, usually as a consequence of a dilatancy-caused pre-pressure drop and an accompanying increase in effective stress. These deformations are accompanied by transient loss of shear resistance rather than permanent loss of shear strength.

Reference: 'Definition of terms related to liquefaction', *ASCE Jl Geotech. Eng. Div.*, **104**, Pt GT9, September 1978.

limit state design The limit state method of design takes into account the interactive effects of the various forces acting on a structure and the structure's ability to resist these forces. Two conditions of failure in a structure are considered: (a) the ultimate limit state and (b) the serviceability limit state. The ultimate limit state corresponds to collapse or failure of a structure due to overstressing or rupture of the component materials, or excessive movement of the structure due to rupture of the supporting soil or soil retained by the structure. The serviceability limit state is that of excessive deformation of the structure. This will be dependent on the required performance of the structure.

linear strain The change in length per unit length in a given direction.

lineation Any linear structure found in a bed of rock. It is usually implied that lineations have a consistent orientation, and may be due to the preferred alignment of certain minerals or be due to the intersection of planar features such as joint or bedding planes.

liquefaction The act or process of transforming any substance into a liquid. In cohesionless soils the transformation is from a solid state to a liquefied state, as a consequence of increased pore pressure and reduced *effective stress*. Liquefaction is thus defined as a changing of state that is independent of the initiating disturbance, which could be a static, vibratory, sea wave, or shock loading, or a change in groundwater pressure. The definition is also independent of deformation of ground failure movements that might follow the transformation. Liquefaction always produces a transient loss of shear resistance but does not always produce a longer-term reduction of shear strength.

Reference: 'Definition of terms related to liquefaction', *ASCE Jl Geotech. Eng. Div.*, **104**, Pt GT9, September 1978.

liquidated damages A contractual term for penalty payment (usually per day or per week) by the contractor to the employer for failure to complete the whole of the works specified in the contract within the prescribed time limit or any extension thereof.

liquidity index (LI) The ratio of the difference between the *moisture content* (*w*) and the *plastic limit (PL)* of a soil to its *plasticity index (PI)*: LI = (*w* − PL)/PI. *See Atterberg limits and soil consistency.*

liquid limit (LL) The moisture content at which a soil passes from the plastic to the liquid state, as determined by the liquid limit test. The preferred method of test is now by use of the *cone penetrometer* rather than the original *Casagrande liquid limit apparatus*, which has been found to give results somewhat dependent on the judgement of the operator and the proper maintenance of the equipment. *See Atterberg limits and soil consistency.*

Reference: BS 1377 : 1975.

lithification The process by which a loose sediment is converted into a sedimentary rock. Lithification is a general term for the penecontemporaneous processes of compression, cementation and growth of secondary minerals. If the process involves more extensive mineralogic changes at elevated temperatures and pressures, it is termed *diagenesis*, and then *metamorphism*.

lithostatic pressure The analogous concept for a rock body to that of hydrostatic pressure in a water body—i.e. that at a given depth the pressure is equal in all directions and has a magnitude equal to the depth times the unit weight of the rock. *See isostacy.*

load factor method A method of defining the factor of safety in design wherein it is applied to the load and used, for example, in pile and bearing capacity problems. *See also permissible stress method; partial factor method.*

load and resistance factor design (LRFD) A design method based on limit states of strength and serviceability instead of the traditional method of using an allowable stress expressed as a percentage of the yield stress. The strength limit relates to the capacity of a structure to survive extreme loads and the serviceability limit state is related to an acceptable performance of the structure during its service life. *See limit state design.*

loading intensity, or unit pressure *Gross loading intensity* (*q*) is the intensity of vertical loading at the base of a foundation due to all loads above that level. *Net loading intensity* (q_n) is the change in intensity of vertical loading resulting from construction. Thus, the net value is the difference between the gross loading intensity before building operations are commenced and the gross loading intensity after the structure is completed and fully loaded. Normally an increase results from construction, but in some cases (e.g. basement construction) a reduction in loading intensity at foundation level is brought about. Note that the loading intensity applied to the ground is unrelated to the ability of the ground to withstand that pressure.

load structures Interpenetrative structures that form at the boundary of two dissimilar materials in response to density or pressure differences. Load structures are often found in soft sediments that are deposited rapidly, so generating excess pore pressures. *See also ball structure.*

loam A term generally taken to mean a soil comprising more or less equal proportions of *sand, silt* and *clay.*

local side friction (f_s) The average unit side friction acting on the friction sleeve in the standard static *cone penetration test (CPT).* *See also Dutch sounding test.*

locked sands A class of sands that occur naturally, composed of predominantly sand-sized particles and characterised by absence of cohesion, high quartzose minerology, high strength, steeply curved failure envelopes, low porosities and considerable geological age.

Reference: 'Locked sand', M. B. Dusseault and N. R. Morgenstern, *Qly Jl Eng. Geol.,* **12**, 117–131, 1979.

lodgement till *See till.*

loess A windblown silt or silty clay characterised by a loose porous structure which can collapse when wetted. *See collapsing soils.*

lognormal distribution A set of values has a lognormal distribution if the frequency of occurrence of the logarithms of the values forms a normal distribution. In geotechnics the main application lies in *particle size* analysis; grains sorted by flowing water have a lognormal distribution of diameters. If grain size is plotted to a logarithmic scale, a normal distribution curve is obtained directly. This accounts for the use of logarithmic graph paper and of the logarithmic *phi (ϕ) scale.*

Loksleeve HD2/Macalloy ground anchor Marketed by CBP (UK) Ltd, this system comprises factory-made units of composite hollow sleeves with inner and outer convoluted sheaths with an annular cementitious or polyester resin grout.

longwall *See subsidence.*

Lorac A medium-range surveying system similar to *Decca navigator system, Toran* and *Raydist.*

142

Loran A long-range pulse-type electronic positioning system maintained by the USA Government.

Los Angeles abrasion test A test developed for highway aggregates, wherein a graded sample is subjected to attrition by wear between rock pieces and also by impact forces produced by an abrasive charge of steel spheres. *See sand blast test.*

losing streams Streams (also called *effluent streams*) that lose water by seepage to the underlying groundwater. The seepage rate depends on the depth of the stream, the *permeability* of the streambed material and underlying soil, and the shape of the stream channel.

Love wave A surface seismic wave characterised by purely horizontal motion perpendicular to the direction of propagation.

Lowe and Karafiath method of slope stability analysis A method developed for analysing dams with either earth or rock cores during conditions of rapid drawdown. It is based on the method of slices, taking into account earth and water forces on the sides of the slices, and involves analysis both at the pre-drawdown and post-drawdown stages.

Reference: 'Stability of earth dams upon drawdown', John Lowe, III and L. Karafiath, *Proc. 1st Pan-American Conf. Soil Mech. and Found Eng.*, **2**, pp. 537–553, Mexico, 1959.

lowest astronomical tide (LAT) The lowest level that can be predicted to occur under average meterological conditions and under any combination of astronomical conditions. It is often the level selected as the datum for soundings on navigational charts.

low oblique (aerial) photograph An aerial photograph which is not vertical, but does not include the line of the horizon. *See also high oblique (aerial) photograph.*

lugeon A unit expressing the water acceptance of one litre per metre per minute obtained during a packer permeability test made in a BX (60.3 mm diameter) borehole at a pressure of $10 \, kg/cm^2$ ($\simeq 10$ atmospheres $\simeq 100$ m water head) over a period of 10 min. 1 lugeon is defined by M. Gignoux and R. Barbeir as being an acceptable leakage rate for dams higher than 30 m and up to 3 lugeons as acceptable for those lower than 30 m. One lugeon is roughly equivalent to a *permeability* of 10^{-7} m/s. However, it should be noted that the test is a rough and ready one, as it departs from the theoretical value of K (*coefficient of permeability*) given by Darcy, since *Darcy's law* is based on *laminar flow*, which is unlikely to occur in fissured rock, where the velocity of flow through the *fissures* is probably relatively high compared to the outflow in the borehole.

Reference: *Géologie des barrages*, M. Gignoux and R. Barbeir, p. 281, Masson, 1955.

L-waves Long-wavelength seismic surface waves produced by *earthquakes.*

lysimeter An instrument that is buried within a soil profile to collect *groundwater*, and so to enable the amount of water flow to be determined. The technique is used at shallow depth in, for example, studies of rainfall and percolation.

M

Mackintosh boring and prospecting tool A portable manually operated soil probing device comprising a string of light rods to which a probing point can be fitted and then driven into the ground to obtain a soil resistance–depth profile, or which can be fitted with a variety of small tools to obtain samples of the ground.

McClelland piston sampler A variation of the *Osterberg piston sampler* whereby a shear pin holds the actuating piston until pressure has been built up to such a degree that the pin shears and causes the piston to move rapidly into the soil.

macroscopic (megascopic) Visible to the naked eye, as opposed to **microscopic** and requiring an optical or electronic aid to detection.

mafic High in magnesia and iron and correspondingly low in silica.

magmatic water *Juvenile water* derived from molten rock material during cooling and crystallisation.

magnetic anomaly A variation from the expected value of the geomagnetic field at a particular location, often as a result of the presence of rocks with remanent magnetisation or of high magnetic susceptibility. Basic igneous rocks and certain orebodies are notable in this respect, as are ferromagnetic artefacts.

magnetic prospecting The measurement of the geomagnetic field with the object of locating *magnetic anomalies.* In geotechnics the technique is often employed to locate artefacts such as steel pipes, reinforced concrete, and so on.

magnetic surveying method The magnetic surveying method depends on measuring local variations in the Earth's total magnetic field strength which are due to changes in magnetic susceptibility of the ground or to the presence of permanently magnetised bodies. It is useful for detecting abandoned mine shafts, basic dyke rocks, submerged wrecks or buried ferrous objects (e.g. pipes). Instruments used include mechanical, *proton* and fluxgate *magnetometers.*

magnetometers Instruments for measuring the strength of the geomagnetic field. Two types are in common field use: the *fluxgate magnetometer*, which operates by electromagnetic induction, and the *proton magnetometer*, in which the gyration of water molecules is utilised. The proton magnetometer is portable and convenient for site use, whereas the fluxgate magnetometer is used primarily during aeromagnetic surveys.

magnitude of earthquake A term describing the total amount of energy released during an earthquake; it is related to the potential damage that can occur. *See earthquakes.*

major principal strain The maximum *strain* corresponding to one of the three orthogonal directions, the shear strains of which are equal to zero. *See also* **average stress, or octahedral normal stress; intermediate principal strain; intermediate (σ_2) principal stress; major (σ_2) principal stress; minor principal strain; minor (σ_3) principal stress; octahedral shear stress.**

major principal stress (σ_1) The major stress acting on one of the three orthogonal planes where shear stresses are equal to zero. *See also* **average stress, or octahedral normal stress; intermediate principal strain; intermediate (σ_2) principal stress; major principal strain; minor principal strain; minor (σ_3) principal stress; octahedral shear stress.**

manometer An instrument used for measuring pressure differences in gases and fluids.

marbled A term describing the coloured patterning of a soil by two or more distinct colours in about equal proportions.

marsh An area of soft ground due to frequent flooding but usually covered by vegetation.

marsh funnel An instrument used to measure the viscosity of drilling fluids (muds) by timing the flow through a standard orifice in the funnel into a calibrated jug.

mass movement A general name for the movement of material downslope under gravity. Mass movements are divided into rapid movements, which include *flow slides* (landslides), *debris flows* and *solifluction*, and slow movements such as *creep*. *See also classification of landslides.*

mass number The nucleon number (A) which is the integer nearest to the atomic mass of an isotope—i.e. the number of nucleons in the nucleus of an atom.

mass properties Certain properties, such as *compressibility*, water transmission (*transmissibility*) and bulk *shear strength* are influenced by jointing in rocks, and thus differ significantly when measured in the field on large rock masses (mass properties) from their corresponding laboratory values measured on small specimens (*material properties*).

mass wasting The loss of material from a slope by any process of *mass movement*.

master joint A member of a joint family which has particular continuity, and which may have influenced the formation and orientation of other joints in its vicinity.

material properties Properties of a rock which are measured on small laboratory specimens rather than in the field. As such they take no account of the influence of structural features, except on the scale of the samples themselves. *See also* ***mass properties***.

maximum and minimum densities *See relative density.*

maximum dry density The dry density obtained by use of a specified amount of compaction at the *optimum moisture content (see Figure A.3)*. *See also* ***dry density of soil; soil compaction***.
    Reference: BS 1377 : 1975.

mean The average value of the points in a distribution. The arithmetic mean is the sum of the values divided by their number; the root mean square is the arithmetic mean of the sum of the squares of the values.

mean high water neaps (MHWN) The average, over a long period of time, of the heights of two successive high waters at neaps.

mean high water springs (MHWS) The average, over a long period of time, of the heights of two successive high waters at springs.

mean low water neaps (MLWN) The average, over a long period of time, of the heights of two successive low waters at neaps.

mean low water springs (MLWS) The average, over a long period of time, of the heights of two successive low waters at springs.

mean sea level (MSL) The average level of the sea surface over a long period of time, preferably 18.6 years (one cycle of the Moon's nodes), or the average level which would exist in the absence of tides.

mechanical analysis The method, usually using hydrometer analysis, by which the *particle size distribution* of the silt and clay fractions of a soil is determined. *See also* ***sieve analysis***.

mechanical composition The *particle size distribution* of the soil.

median The central value of a distribution, above and below which lie half the number of values, respectively.

melt-out till *See till.*

Menard pressure meter The Menard pressure meter comprises a cylindrical probe with a membrane which can be pressurised and expanded against the sides of a predrilled hole. A knowledge of the pressure required to cause a change in diameter of the hole is used to determine the stress–strain parameters of the ground and thus provide *bearing capacity* and *settlement* information for foundation design. *See also* ***dynamic consolidation***.

meniscus The cup-shaped upper surface that a fluid assumes in a capillary tube due to surface tension of the fluid and the attraction between the fluid and the container. *See also* ***capillarity***.

Mercalli (modified) scale of earthquake intensity A scale of earthquake intensity introduced in 1931 and given in *Table 18*. *See earthquakes.*

mercury barometer *See barometer.*

metamorphic water *Juvenile water* driven out of rocks during metamorphism.

metamorthism *See lithification.*

meteoric water *Juvenile water* derived from outer space.

methane A colourless, odourless, tasteless, combustible gas. It is found in some groundwaters, escape from which may produce an explosive atmosphere in air if a percentage within the range 5–15 is reached. Methane is also produced from the breakdown of organic matter in an aerobic condition, owing to the existence of bacteria in the presence of free oxygen.

metre The basic metric unit of length in the International System (SI); defined as 1 650 763.73 vacuum wavelengths of the orange line emitted by the krypton isotope 86. Originally one ten-millionth part of the distance between the pole and the equator along the meridian of the Paris Observatory.

micro A prefix denoting one-millionth (10^{-6}).

micrometre (μm) A unit of length equal to 10^{-3} mm.

micropiles Small-diameter piles (60–300 mm) which can be installed up to a 15° rake by drilling with rotary cutting tools capable of penetrating through concrete, masonry, rock or soil layers, and comprising steel tubing which is grouted up with a high-strength cement–sand mix injected under pressure. They are used for underpinning damaged buildings and strengthening structures such as retaining walls, river banks and embankments.

microscopic Invisible to the naked eye and requiring an optical or electronic aid to detection. *See also* **macroscopic (megascopic)**.

microseisms Faint earth tremors due to natural causes such as wind, water waves, etc. Instruments sensitive to such movements are used to detect rock and earth slides and for monitoring faults for potential *earthquakes.*

microsonic activity Microactivity (subaudible rock noise—SARN) occurs during incipient failure conditions of rock and soil slopes caused by slight (*creep*-like) movement of the slope mass and consequent shear of the rock or soil particles as the slope deforms. Ultra-sensitive listening devices can be used to monitor and record these subsurface microseismic and microacoustic transient disturbances by placing sensitive electromagnetic geophones in the ground at various levels. The acoustic activity can then be evaluated to determine the probable spatial location of the sliding zone(s). Research to date indicates that acoustic activity tends to provide a qualitative rather than a quantitative assessment of stability with an increase in SARN rate reflecting a decrease in stability, and *vice versa.*

Miga pile   *See Franki piling systems.*

milligal   One-thousanth of a *gal*, an acceleration of 10^{-3} cm s^{-2}. Milligals are the units used to measure an *anomaly* in the Earth's gravity field.

MINI-FIX   A miniaturised *HI-FIX* system intended primarily for use at short ranges in operations requiring accuracies of the order of one or two metres. The equipment can be set up in vessels or on vehicles and can be used for projects such as inshore hydrographic surveys, oil prospecting control or local dredging operations. *See Decca Navigation System; SEA-FIX.*

mini-ranger/trisponder   An electronic marine survey instrument comprising a range console (a small receiver–transmitter unit) and an omnidirectional antenna aboard the survey vessel, and two or more reference stations (trisponders) which are located at co-ordinated shore stations. In operation, it measures distances from the boat to each reference station which are continuously displayed by the console. By using prepared *circle (lattice) charts* the vessel can be positioned. The mini-ranger MRS III system is integrated with a desk-top computer and a flat-bed plotter which allow of rapid automatic plotting of ranges from the reference stations on to a prepared rectangular grid. This allows the surveyor to concentrate on steering the vessel and monitoring other instruments.

mini-tunnel system   A tunnelling system with diameters ranging from 1 m to 1.3 m and constructed using three identical unreinforced concrete segments per tunnel ring, each of 600 mm length. Each ring is built within the rear section of the shield by use of a purpose-made erector. The shield is jacked forward off the concrete segmental lining cutting an overbreak which is filled with pea gravel, to ensure immediate ground support for the lining. After excavation of the tunnel, the pea gravel is grouted to seal the tunnel. Applications include small utility tunnels, pilot bores and sewage networks.

minor principal strain   The minimum *strain* corresponding to one of the three orthogonal directions, the shear strains of which are equal to zero. *See also average stress, or octahedral normal stress; intermediate principal strain; intermediate (σ_2) principal stress; major principal strain; major (σ_1) principal stress; minor (σ_3) principal stress; octahedral shear stress.*

minor principal stress (σ_3)   The minimum *stress* acting on one of the three orthogonal planes where shear stresses are equal to zero. *See also average stress, or octahedral normal stress; intermediate principal strain; intermediate (σ_2) principal stress; major principal* *strain; major* (σ_1) *principal stress; minor principal strain; octahedral shear stress.*

Mintrop wave   *See head wave.*

mirror stereoscope   An instrument for binocular observation of stereoscopic pairs of photographs.

Mo   A Scandinavian term for the size of soil particle intermediate between *silt* and fine *sand.*

mode   The modal class in a frequency distribution is that class with the greatest number of entries. Thus, the mode is the value or interval of values that forms that class.

modified AASHO compaction test   *See modified Proctor compaction test.*

modified Dorry test   *See Dorry test.*

modified Mercalli (MM) scale   *See earthquakes; Table 18.*

modified Proctor compaction test   A modification of the *standard Proctor compaction test* whereby a 10 lb rammer and an 18 in drop is used to give 56 250 ft-lb per ft^3 of energy. It is also known as the *modified AASHO compaction test* (T1-80); ASTM methods D1557-70 A, B, C, D; and BS Compaction Test (10 lb (4.5 kg) rammer method; BS 1377 : 1975 Test 13).

modulus of compressibility   The ratio between an isotropic stress change and the corresponding volume change per unit volume.

modulus of deformation   A term synonymous with *secant modulus of elasticity.*

modulus of elasticity, or Young's modulus   The ratio between a linear *stress* σ and the corresponding linear *strain* ε, this being a constant for a perfectly elastic material. The ratio μ, between the linear stress σ and the corresponding linear strain ε_e measured at right angles to the direction of the stress is known as *Poisson's ratio. See stress–strain relationships.*

modulus of linear deformation   The ratio between a given normal stress change and the linear strain change in the same direction (all other stresses being constant).

modulus of shear deformation   The ratio between a given shear stress change and the corresponding shear strain change (all other stresses being constant). *See stress–strain relationships.*

modulus of subgrade reaction   The modulus (or coefficient) of subgrade reaction is the constant of proportionality obtained by Westergaad in his theory of the stresses and deflections in concrete slabs in which the elastic reaction of the subgrade against the slab is assumed to be vertical and proportional at all points to the vertical deflection. This value *(k)* is measured in force/length^3 units (e.g. $MN\,m^{-2}\,m^{-1}$ or $lb\,in^{-2}\,in^{-1}$) and is used to determine data for the design of concrete pavements and foundation rafts. The modulus is obtained from *in situ* **plate bearing tests**, the test procedure being set

149

out in *Soil Mechanics for Road Engineers* (HMSO, London, 1955). Approximate values of vertical and horizontal subgrade reaction in both sand and clay soils are given by Terzaghi in *Géotechnique*, V, No. 4, 1955. Also, H. F. Winterkorn and H. Y. Fang present, in *Foundation Engineering Handbook*, pp. 516–517, 1975, various methods for obtaining the modulus of subgrade reaction, including *California Bearing Ratio (CBR) tests; consolidation tests; plate bearing tests; triaxial compression tests;* and *unconfined compression tests.*

Mohr diagram, or Mohr circle A graphical construction on which normal stresses (X-axis) are plotted against shear stresses (Y-axis) such that any point in a circle (Mohr circle) centred on the X-axis represents the co-ordinate of stress on a particular plane. The results of a triaxial test may be plotted on such a diagram so that a circle having a diameter ($\sigma_1 - \sigma_3$) cuts the X-axis at abscissae of σ_1 (*deviator stress*) and σ_3 (cell pressure). The inclination of a line tangentially cutting a number of such circles (usually three) gives the *angle of shearing resistance* (ϕ) of the material and the *cohesion* intercept *(c)* where it cuts the Y-axis.

Moh scale of hardness A scale of relative hardness which is graduated from 1 (soft) to 10 (hard). Cardinal points on the scale are defined by the hardnesses of talc (1), gypsum (2), calcite (3), fluorspar (4), apatite (5), orthoclase feldspar (6), quartz (7), topaz (8), corundum (9) and diamond (10). Any mineral can be placed on the scale by its relative hardness—i.e. by its ability to scratch or to be scratched by other minerals of known hardness. Moh hardness provides a convenient criterion for the field identification of minerals—for example, calcite (4) is distinguished from quartz (7) by the inability of steel blade (6) to scratch the latter.

moisture condition value (MCV) test A test developed by the British Transport and Road Research Laboratory (TRRL) as a construction control method for assessing the suitability of earthwork materials to be used in embankments. The test involves measuring the compactive effort required to fully compact a soil in a 100 mm diameter mould with a 7 kg mass rammer dropping freely through a height (adjustable) of 250 mm. A curve is plotted for the number of blows delivered against change in penetration, from which the MCV is defined as 10 times the logarithm to the base 10 of the number of blows corresponding to a change in penetration of 5 mm. The test avoids the normal delays associated with the need for moisture content tests using standard compaction tests, and the MCV has been shown to be related to the performance of earthmoving equipment and to various physical parameters such as undrained *shear strength* of remoulded soil, *California Bearing Ratio (CBR)* and *liquidity index.*

Reference: *The Rapid Measurement of the Moisture Condition of Earthwork Material*, A. W. Parsons, UK TRRL Report 750, 1976.

moisture content Defined in *soil mechanics* as the loss in weight of a soil expressed as a percentage of the dry material when it is dried to constant weight at 105° C.

moisture–density curves *See standard Proctor compaction test.*

Monodur process A technique used in *grouting*, involving a single injection process with dilute sodium silicate and organic chemicals forming acids (Monodur solution). The solution has a low viscosity (about 3 cP) and can harden soils to give compressive strengths up to about $10\,000\,kN/m^2$. Often used in conjunction with the *Joosten process*.

monolith An open *caisson* of heavy mass concrete or masonry construction, containing one or more cells for excavation.
Reference: CP 2004 : 1972.

monthly certificate An interim certificate approved and issued by the engineer for a project, based on the *monthly statement* submitted by the contractor for work executed and goods, materials and services supplied by him up to the end of that month.

monthly statement A statement submitted to the engineer for a project, showing the estimated value of the permanent works executed and details of any goods, materials or services provided up to the end of that month for which the contractor feels he is entitled to receive an interim payment.

moonpool A central well in a vessel to allow drilling to take place through the body of a ship into the sea-bed or river-bed.

Moore free-fall corer A type of *drop sampler, or drop corer* which can be used, mainly in deep water, without a winch and cable. The core barrel is attached to a buoyant chamber which is released on impact with the sea-bed, pulling the full core tube back to the surface.

Morgenstern method of slope stability analysis A method based on Bishop and Morgenstern's method of slices. It is mainly used to determine the factor of safety of earth slopes during *rapid drawdown*, as in the case when the water level in a reservoir is lowered. A series of charts give values of factor of safety against the drawdown ratio L/H over a range of ϕ'-values for various values of slope angle β and $c'/\gamma H$, where: L/H = the amount of drawdown divided by the original height of the slope; ϕ' = the effective angle of shearing resistance; c' = the effective cohesion; γ = the soil density; and H = the height of the slope.
Reference: 'Stability charts for earth slopes during rapid drawdown', N. Morgenstern, *Géotechnique*, **xiii**, No. 2, pp. 121–131, 1963.

morphologic unit A concept applied in *terrain analysis*. A morphologic unit is a recurrent land form or group of land forms that is formed by a particular genetic process.

MOSS Computer program acronym for MOdelling SystemS, marketed by a consortium of Durham, Northamptonshire and West Sussex County Councils, UK, for three-dimensional modelling exercises including waste disposal, land reclamation and hydrographic studies. The system can be used to prepare maps from land survey information and to record the location of gas, water and electricity services.

motorway A road specifically designed for motor traffic, with complete grade separation and limited access points. Equivalent terms in use include express-ways, autostradas, autobahns, etc.

mottled A term describing the coloured patterning of small areas of soils by one or more colours in another predominant colour.

MSK scale of earthquake intensity A scale of earthquake intensity introduced by Drs Medvedev, Sponheuer and Karnik. *See earthquakes.*

muck soil An American term for black decomposed material containing some fibrous material and a high percentage of mineral matter.

mud (1) A mixture of clay particles and water used in *rotary drilling*. *See circulation.* (2) Admixtures of silt and clay in an almost fluid state. Mudflats comprise areas of mud periodically covered by floods or tides.

mud balance An instrument used to determine the density of drilling fluid (*mud*), wherein a weight can be moved along a pivoted calibrated arm to balance a fixed volume of the fluid.

mud cake Mud deposited on the walls of a *borehole* as it loses water into porous strata. Its low *permeability* tends to reduce further loss of fluid into the formation.

mudflows This type of flow occurs in clay beds containing pockets, layers or laminations of *silt* and *sand*, e.g. *varved clays* and *glacial deposits*, when high pore pressures are generated by heavy rainfalls. Such flows generally require a ground inclination of between 5° and 15° for generation and have been classified by Skempton and Hutchinson as 'elongate', where the length-to-width ratio is greater than 10, or 'lobate' where the L/B ratio is less than 10. *See classification of landslides.*

Reference: 'Stability of natural slopes and embankment foundations', A. W. Skempton and J. N. Hutchinson, *Proc. 7th Int. Conf. Soil Mech. and Found. Eng.*, Mexico, 1969.

mullion structures Rod-like linear structures which develop in competent rocks as a result of *compression*. Mullions are often associated with folds, and are orientated perpendicular to the direction of greatest stress, so lying parallel to the fold axes.

multi-stage triaxial compression test An undrained test carried out on a single specimen but with (usually three) different lateral pressures. Its main use is where only one specimen can be prepared from each standard sampling tube.

multi-wheel roller *See pneumatic-tyred roller.*

multivariate Depending on, or correlated with, two or more independent variables. In statistics multivariate techniques refer to methods of projecting sets of dependent variables on to spaces of independent variables, by use of some criterion of best fit such as *least squares* (*method of*). Well-known techniques of this kind include *factor analysis* and *principal components analysis*, as well as the *fitting of trend surfaces*.

Munsell charts *See Munsell colour.*

Munsell colour A standard measurement of colour, usually of a soil, based on a comparison with a set of graded colour charts (*Munsell charts*).

N

nano Prefix denoting one thousand millionth (10^{-9}).

narrow strip foundations *See strip foundation.*

natural frequency of soil-foundation system The critical or resonant frequency of an entire system comprising a machine, its foundation and supporting soil mass; it is the frequency at which the system will vibrate at maximum amplitude when the machine is run at the same frequency. Where foundation soils are susceptible to vibration and likely to compact and settle (e.g. loose sand) owing to this cause, special precautions need to be taken in the design of the machine mounting and foundation to damp the system down, if the machine is to be run at speeds likely to cause resonance.

natural moisture content Moisture content in the natural state of a material.

NAVSTAR An American military satellite navigation system relying on 24 satellites circling the Earth every 12 h. The accurate location of terrestrial vehicles such as small ships or even cars is possible by their carriage of lightweight radio transmitters.

NAVTRAK Doppler navigator An electronic navigational aid to allow a vessel to follow a track in water depths of about 200 m, basically comprising transducers which emit narrow-beam signals fore, aft, port and starboard, and equipment to translate the received echos to determine a safe route.

NCB cone indenter A portable instrument developed by the Mining Research and Development Establishment (MRDE) of the British National Coal Board (NCB) for determining rock strength without requiring accurately shaped and finished specimens, by measuring its resistance to indentation by a hardened tungsten carbide 40° apex cone:

$$\text{cone indenter hardness value } I = \frac{D}{P}$$

where D = nominal deflection of a steel strip forming part of the equipment; P = penetration of the specimen by the cone.
Reference: MRDE Handbook No. 5, *NCB Cone Indenter*, 1977.

neap tides The two occasions in a cycle of one lunar month (approximately $29\frac{1}{2}$ days) when the average range of two successive tides is least.

near-infra-red That part of the *electromagnetic spectrum* between visible light and thermal infra-red.

necking, or waisting, of piles The reduction in cross-sectional area of a pile shaft that can occur where poor construction techniques allow the inflow of soft or loose water-bearing soils to squeeze out the freshly placed concrete forming the shaft.

needle A demolition term for a member inserted into or through a wall to give it temporary or permanent support. A needle used in conjunction with a *flying* (or *raking*) *shore* consists of a short member passing through the vertical wall piece and into the wall to form the abutment to the shore.
Reference: CP 2004 : 1972.

negative skin friction (piling) The downward force transmitted to a pile by the surrounding ground where such material can settle relative to the soil/rock surrounding the lower part of the pile—e.g. where a pile is driven through loose fill material to a competent bearing stratum, with the fill still settling under its own weight. *See also skin friction (piling)*.

net loading intensity The change in intensity of vertical loading at the base of a foundation resulting from construction. Thus, the net value is the difference between the *gross loading intensity* before building operations are commenced and the gross loading intensity after the structure is completed and fully loaded. Normally an increase results from construction, but in some works a reduction in loading intensity at foundation level is brought about, for example, where the weight of material removed for a basement construction is in excess of the subsequent superstructural loads applied. Note that the loading intensity applied to the ground is unrelated to the ability of the ground to withstand that pressure. *See also allowable bearing pressure; loading intensity, or unit pressure*.

n per cent diameter (D_n) The diameter corresponding to a percentage by weight of finer particles (i.e. particles of smaller diameter).

net positive suction head (NPSH) A term defining the head required to cause water to flow through the suction pipe of a pump and finally enter the eye of the impeller; it is obtained from the atmospheric pressure with or without any static head. The required NPSH is a function of the pump design, whereas the available NPSH is a function of the system in which the pump operates. To operate successfully, the available NPSH of a pump installation must be at least equal to the NPSH of the pump. The NPSH is equal to the barometric pressure, minus the friction losses in suction pumping minus the vapour pressure of the liquid (plus the static head on suction where the source of liquid is above the pump or minus the static head where the source of liquid is below the pump).

net safe bearing capacity *See bearing capacity.*

net ultimate bearing capacity *See bearing capacity.*

neutral stress, pressure An alternative name for the spherical component of a stress system. In geotechnics the term is often used for the pore water pressure.

neutral surface A surface of no strain, found within a bed of rock that has been folded as a simple beam. The neutral surface may be identified physically from the fact that geological objects within the bed are undeformed along that surface.

neutron-lifetime log The neutron-lifetime log is obtained by down-hole radiation logging by counting the thermal neutrons over discrete time intervals following periodic releases of neutrons into the surrounding formations, from a van de Graaf generator in the sonde. It is similar to the *thermal decay-time log*, wherein measurements are made during thin discrete time intervals of the gamma rays that result from the capture of neutrons by nuclei as the neutrons lose energy in collisions following their release into the ground. The logs may be plots of either the reciprocal of the percentage of thermal neutrons that decay in unit time (thermal decay-time) or the time required for the number of thermal neutrons to fall to half value (neutron-lifetime). *See also chlorine log.*

neutron log A log obtainable in either a cased or an uncased well or borehole by lowering a sonde containing a source of fast electrons and a detector for measuring the gamma-ray radiation produced when the neutrons are captured by atomic nuclei, after being slowed to thermal speed by collision with mainly hydrogen atoms in the surrounding ground. Since the gamma-ray radiation is proportional to the number of hydrogen atoms present, the hydrogen density will indicate the proportion of moisture present in the formation and, hence, its *porosity*. The detector can record *gamma rays*, thermal neutrons or epithermal neutrons (i.e. those just above thermal speed). *See also neutron moisture gauge; side wall neutron log.*

neutron moisture gauge An instrument comprising a neutron source and detector, an electronic supply and an indicating unit. By measuring the flux of slow neutrons near the detector, the hydrogen content of the medium can be determined and, hence, the water (H_2O) content. Common neutron sources used include radium and americium.

neutron–natural gamma log A *borehole record* obtained with a probe containing a radioactive source such as americium or beryllium, a neutron detector for density determination and either a scintillation- or Geiger–Müller-type *gamma-ray* detector for *moisture content* determination.

Newlyn datum *See ordnance datum.*

Newmark's charts A graphical method using influence charts for computing the vertical pressure in the soil beneath a loaded area. In the Newmark chart *(Figure N.1)* the loaded area is drawn to a scale such that the depth at which it is required to compute the vertical pressure is made equal to the distance AB. The scaled drawing is then placed on the chart with the spatial location of the required vertical pressure made coincident with the centre of the

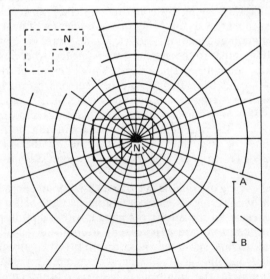

Figure N.1 Newmark chart for determining intensity of vertical pressure below a point N of a structure

concentric circles. The pressure is then calculated by summation of the 'influence squares' lying within the scaled area.

Reference: *Soil Mechanics in Engineering Practice*, K. Terzaghi and R. B. Peck, Wiley, 1948.

newton A unit of force, being the force which is exerted by a mass of 1 kg when subjected to an acceleration of 1 m/s^2. It may be noted that the gravitational unit of force, kgf, is that force exerted by a mass of 1 kg when subjected to standard gravitational acceleration of $9.806\,65 \text{ m/s}^2$—i.e. $1 \text{ kgf} = 9.806\,65 \text{ kg m/s}^2 = 9.806\,65 \text{ N}$.

Newtonian fluid A fluid the shearing deformation of which is directly proportional to the applied stress level. The viscosity (μ) of such a fluid is defined by the slope of the line having the equation (*see Figure N.2*)

$$y = \mu \frac{\mathrm{d}v}{\mathrm{d}z}$$

A non-Newtonian fluid is characterised by an upward concave curve (shown dotted in *Figure N.2*). This is due to the presence of particles in the fluid (e.g. as in a clay) and the fact that the viscosity depends upon the rate at which shearing takes place.

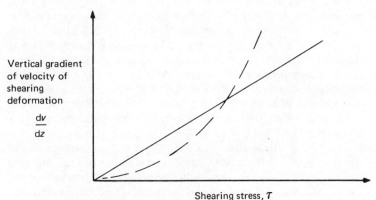

Vertical gradient of velocity of shearing deformation

$$\frac{\mathrm{d}v}{\mathrm{d}z}$$

Shearing stress, T

Figure N.2   Curve of a newtonian fluid

NHBC National House Building Council.

noise A form of acoustic energy often described as unwanted *sound*. *See also decibel; equivalent continuous sound level; reflected sound; sound attenuation; sound level.*

nomograph A chart, generally comprising a number of straight or curved graduated lines, whereby the use of a straight-edge across them allows of the rapid calculation of formulae.

non-affine Descriptive of a pattern of deformation in which originally parallel elements do not remain parallel during deformation. This is also known as *non-homogeneous deformation*. *See also* **affine**.

non-cohesive soils Non-plastic soils comprising coarse, mainly siliceous and unaltered fragments of rock weathering (e.g. *sand* and *gravel*).

non-equilibrium well formulae *See Theis method.*

non-frost-susceptible materials Cohesionless materials with generally 10 per cent or less of material passing a 75 μm sieve, such as crushed rock, gravel, sand, slag and cinders, in which significant detrimental ice segregation does not occur under normal freezing conditions. Cohesive soils are regarded as non-frost-susceptible when the *plasticity index (PI)* is greater than 15 per cent for well-drained soils, or 20 per cent for poorly drained soils (i.e. when the water table is within 600 mm of the formation level). *See frost-susceptible soil; frost susceptibility test.*

non-homogeneous *See non-affine.*

non-plastic Descriptive of a soil with a *plasticity* of zero or one on which the *plastic limit (PL)* cannot be determined.
    Reference: BS 1377 : 1975.

normal distribution A statistical distribution of values that occurs when a large number of observations are made of the same property, but with each observation subject to a random error. The frequency graph of a normal distribution is bell-shaped, and is centred on the mean value of the observations. The importance of the normal distribution arises from the fact that it occurs in very many circumstances that involve random events, and also in that the proportion of the observations that lie within a given distance of the mean can be stated exactly in terms of the standard deviation of the distribution. This property forms the basis of many methods of sampling and hypothesis testing from large populations.

normal strain Normal strain is defined in relation to *liquefaction* as the change of length per unit length in a given direction. The term can be applied to deformation in cohesionless soils in either the solid or liquefied state (as may be the case if subject to liquefaction). Strains may be expressed in terms of a dimensionless ratio or as a percentage.
    Reference: 'Definition of terms related to liquefaction', *ASCE Jl Geotech. Eng. Div.*, **104**, Pt GT9, September 1978.

normally consolidated A clay is said to be normally consolidated if it has never been under a pressure greater than the existing *effective overburden pressure*.

nuclear-magnetism log Abbreviated to NML and also called *free-fluid log*; a measurement of a signal of the radiofrequency band emitted by the precession of protons (hydrogen nuclei) in the

Earth's magnetic field as they slowly return from an induced magnetic alignment in an impressed magnetic field. The signal amplitude is measured as a *free-fluid index (FFI)* which provides information on the nature of the free fluid in the surrounding formation—i.e. whether it is gas, water or oil.

nuclear methods for field density determinations *See field density tests.*

number of sets of discontinuities One of the ten parameters selected to describe discontinuities in rock masses, being the number of *joint sets* comprising the intersecting joint system. The rock mass may be further divided by individual discontinuities.

Reference: International Society for Rock Mechanics. Commission on Standardisation of Laboratory and Field Tests, 1977.

Nu-well A chemical in pellet form produced by the Johnson Division of Universal Oil Products (UOP) for treating wells clogged by certain types of incrustants such as lime scale deposits of calcium and magnesium and iron oxide.

N-value The standardised 'blow count' obtained in a *standard penetration test.*

O

^{18}O *See oxygen-18 (^{18}O).*

octahedral shear stress Equal to

$$\frac{\sqrt{(\sigma_1 - \sigma_2)^2 + (\sigma_2 - \sigma_3)^2 + (\sigma_3 - \sigma_1)^2}}{3}$$

where σ_1 = the *major principal stress*; σ_2 = the *intermediate principal stress*; and σ_3 = the *minor principal stress. See also average stress or octahedral stress; intermediate principal strain; major principal strain; minor principal strain.*

oedometer Laboratory apparatus with which the *consolidation* characteristics of small soil samples may be obtained for a range of static loadings and thereby provide information for the analysis of *settlement.*

oedometer modulus (E_{oed}) The reciprocal of the coefficient of volume change ($1/mv$) obtained from a consolidation test on a soil sample. *See coefficient of volume change or compressibility (m_v).*

off-line A computer term for that part of a computer that is not under the control of the *central processing unit (CPU).*

Omega A long-range very-low-frequency all-weather electronic position system developed by the US Navy.

on-line A computer term for the computer section under the control of the *central processing unit (CPU)*.

open caisson *See caissons.*

open hole An uncased *borehole*.

open-hole drilling A method of boring whereby boreholes are sunk without the use of *casing*. *See rotary probe drilling*.

open-jacket platforms Open-jacket type platforms for off-shore gas/oil units or drilling decks and similar uses are normally of open-lattice steelwork construction and designed to transmit the platform loads and environmental forces due to the sea and wind down into the sea-bed by piling. The latter is often time-consuming, especially in rough open-water conditions. *See also gravity-platforms*.

open sheeting Sheeting used to protect the sides of an excavation in reasonably firm ground which is unlikely to slough. Such sheeting normally comprises either vertical poling boards spaced at intervals dependent on the firmness of the ground and supported by *walings* and *struts*, or open-spaced horizontal sheeting held in position by *soldiers* and struts.

optimum moisture content The moisture content at which a specified amount of compaction will produce the *maximum dry density* *(see Figure A.3)*.
    Reference: BS 1377 : 1975.

ordnance bench mark *See bench mark.*

ordnance datum The standard datum for the United Kingdom. Originally assumed as mean sea level at Liverpool but now taken as mean sea level at Newlyn in Cornwall and obtained by taking hourly observations from 1915 to 1921.

Ordnance Survey maps The Ordnance Survey of Great Britain publishes the following maps of interest to the geotechnical engineer and geologist:

1:1250 scale topographical maps covering major urban areas.
1:2500 scale topographical maps covering the majority of Great Britain.
1:10 560 scale (6 inches to 1 mile) topographical maps cover the whole of Great Britain but are being replaced by 1:10 000 scale.
1:50 000 scale first and second series topographical maps covering Great Britain, including the Isle of Man, which are replacing the 1 inch maps.
1:250 000 scale ($\frac{1}{4}$ inch topographical maps).
1:625 000 scale giving specialist information on, for example, route planning, administration areas, archeology and geology.
1:63 360 scale (1 inch) geological maps—being replaced by 1:50 000 scale.

1:10 560 scale (6 inch) geological maps—being replaced by 1:10 000 scale.

1:25 000 scale (about $2\frac{1}{2}$ inches to 1 mile) geological maps of special areas.

1:63 360 and 1:25 000 scale soil maps of Great Britain.

In addition, early editions of Ordnance Survey maps and plans can be viewed at the Map Room, British Museum, London; The National Library of Scotland, in Edinburgh; and The National Museum of Wales, in Aberystwyth.

orientation analysis The analysis of the disposition of the linear or planar elements of a rock fabric, such as the long axes of pebbles, poles to joint surfaces, and so on. The data are normally presented on rose diagrams or stereograms, and may be characterised by values of mean vector(s) and the radius of the circle of dispersion. It is possible to test the dispersion statistically against that expected from a random distribution, and so to evaluate the probability that the data are clustered. It should be noted that there are some objections to the use of circular statistics, and so the results are not always accepted uncritically.

orientation of discontinuity One of ten parameters selected to describe discontinuities in rock masses; it is the attitude of discontinuity in space and is described by the dip direction (azimuth) and dip of the line of steepest declination in the plane of the discontinuity.

Reference: International Society for Rock Mechanics, Commission on Standardisation of Laboratory and Field Tests, 1977.

organic polymers *Hydrophilic* colloids with a strong affinity for water, used in drilling fluids.

Osterberg piston sampler A fixed piston type of sampler for obtaining samples of soft clay and sand. Basically it comprises a thin-walled sample tube (63.5 mm, 76.2 mm or 127 mm diameter) which can be pushed into the soil at the base of a borehole by pumping water down the drill string to a piston on the sampling tube. The system contains an arrangement to prevent overdriving. *See McClelland piston sampler.*

outside clearance ratio (of a soil sampler) A ratio equal to

$$\frac{D-d}{d}$$

where D = the outer diameter of the cutting nose and d = the outer diameter of the barrel shaft.

overbreak The unintended breaking away of rock beyond the desired line of an underground excavation. Overbreak is frequently related to the presence of *joints* or other *discontinuities* which enable the rock to loosen and collapse beyond the design area. Overbreak may be limited by the judicious placement and timing of explosives, and also by preliminary drilling to the line of the excavation in order to focus the effect of the charges.

overburden A term generally used to describe soil or other material overlying the solid (rock) deposits. It is also used to describe material which has to be removed and rejected in order to work a mineral deposit.

overburden pressure The stress at a point in a body of soil or rock that results from the superincumbent load of the soil or rock column above that point. The usual convention is to refer to the vertical normal pressure, which may be in terms of either total or effective stresses.

overconsolidated The state of consolidation of a soil due to the existing effective overburden pressure being less than that which existed at some time in its history. This is a common occurrence where soils have been fully consolidated under strata which have been subsequently removed to some degree by erosion (e.g. by glaciation). Sedimentary rocks such as mudstones and siltstones with little or no existing overburden are clear examples of soils which at one time in their history were subjected to considerable overburden pressure.

overconsolidation ratio (OCR)

$$OCR = \frac{\sigma'_p}{\sigma'_o}$$

where σ'_p = the preconsolidation pressure and σ'_o = the present overburden pressure.

overpumping The removal of water from an *aquifer* at a rate greater than the natural rate of recharge, so leading to a fall in the *groundwater table* or *piezometric surface*.

owner In American contract law, a public body or authority, corporation, association, partnership or individual for whom the *work* is to be performed.

Reference: *Manual of Water Well Construction Practice*, US Environmental Protection Agency, 570/9-75-001.

oxidation–reduction potential *See redox potential*.

oxygen-18 (^{18}O) The heavy stable isotope of oxygen.

oxygen (dissolved) A gas normally comprising about 20 per cent of the Earth's atmosphere, present in all rain and surface water owing to their contact with the atmosphere. The amount present in *groundwater* tends to decrease with depth and may be absent in supplies from deep wells.

ozone   A potent germicide used as an oxidising agent to destroy organic compounds that can produce obnoxious tastes and odours in well-water.

P

packer method of grouting   *See curtain grouting.*

packers   Mechanically or hydraulically operated expanders, such as expanding rubber rings or cup leathers, which are placed in boreholes to confine water or grout, pumped down the hole, to within specified depths.

packer tests   Tests carried out in unlined boreholes sunk in rock formations using expanding packers to isolate a section of the borehole. *See pumping-in permeability tests.*

packing, for piling   A pad of resilient material contained between the helmet and the top of a reinforced concrete pile to minimise damage to the head during driving.
Reference: CP 2004 : 1972.

pad foundation   An isolated foundation to spread a concentrated load.
Reference: CP 2004 : 1972.

paddy-field soil   A loam containing a higher than usual sand content (about 40 per cent) and a correspondingly lower clay content (about 20 per cent).

PAFEC   A computer program for analysis of a wide range of engineering problems by use of finite element methods.

page   A small wooden wedge used in timbering.

Pali Radice Pile   A proprietary system of Fondedile Foundations Ltd, comprising the drilling of small-diameter holes by *rotary drilling* through the foundation structure into the underlying soil and casting *in situ* reinforced concrete in the boreholes.

Parez hydraulic cone/friction sleeve   *See penetrometer (apparatus).*

partial factor method   A method of defining the factor of safety in design wherein separate factors of safety are applied to both the material strength and the load. They are smaller in magnitude than those used in the *permissible stress method* and the *load factor method.*

particle size   Particle size is defined in all branches of geotechnical and soil sciences as the *effective diameter* of the soil particle, which is the minimum sieve opening that allows the particle to pass through. The following table sets out the particle sizes of soils according to the most well known and widely used gradation and classification systems.

Particle Range and Size (mm)	Classification system*				
	BSCS	ASTM	MIT	USDA	ISSS
Boulders	>200	—	—	—	—
Cobbles	60–200	—	—	—	—
Gravel	2.0–60	>2.0	2–185	>1.0	>2.0
Sand	0.06–2.0	0.074–2.0	0.06–2.0	0.05–1.0	0.02–2.0
Silt	0.002–0.06	0.005–0.074	0.002–0.06	0.002–0.05	0.002–0.02
Clay	<0.002	<0.005	<0.002	<0.002	<0.002

* BSCS: British Soil Classification System for Engineering Use.
ASTM: American Society for Testing Materials Classification.
MIT: Massachusetts Institute of Technology Classification.
USDA: United States Department of Agriculture Bureau of Soils Classification.
ISSS: International Society of Soil Sciences Classification.

particle size classification systems Various systems have been evolved for identifying particle size. Identification of soils according to particle (grain) size is normally performed by *sieve analysis* of the coarser material and by hydrometer analysis (*see mechanical analysis*) of the finer materials, or by a combination of both methods. The classification systems generally refer to *gravel, sand, silt* and *clay* for particular ranges of particle size; however, the physical boundaries of such sizes vary between the different systems. *See also particle size.*

particle size distribution The percentages of the various grain sizes present in a soil as determined by sieving and sedimentation. *See mechanical analysis; sieve analysis.*
Reference: BS 1377 : 1975.

pascal A unit of pressure denoted by Pa and equal to 1 newton per square metre (N/m^2) and approximately equal to 1.45×10^{-4} pounds-force per square inch (lbf/in^2). Also 1 bar $= 10^5$ N/m^2 $= 100\,kPa$.

passive earth pressure The upper limiting value of lateral pressure or resistance of a soil on the face of a retaining structure, reached when the structure yields and causes the soil in front of it to be compressed in a horizontal direction. The *coefficient of passive earth pressure* or resistance, K_p, is equal to $(1 + \sin \phi)/(1 - \sin \phi)$, where ϕ is the *angle of internal friction* of the soil.

passivity The state of the surface of a corrodible metal or alloy at which its electrochemical behaviour becomes that of a less reactive metal and its corrosion rate is reduced. This is generally the result of protective film formation.
Reference: CP 1021 : 1979.

pavement (1) A term embracing all forms of carriageway and runway design (*see pavement design*). (2) The usual term in Britain for the pedestrian footpath adjacent to the carriageway in a road. (3) A hard floor surfacing of concrete, wood blocks, tiles, bricks, etc. (term mainly used in the USA).

pavement design Pavements may be designed as either rigid or flexible, the particular method generally being governed by economic and operational requirements. The pavement comprises *sub-base*, base (*roadbase*) and *wearing course*, the combined thicknesses of which must be sufficient to transmit the traffic loads down to the ground (*subgrade*) without causing it distress. Road design is generally based on the ***California Bearing Ratio (CBR) test***, and a long study of the performance of British roads has led to the design methods given in *Road Note 29* (3rd edn), issued by the Department of the Environment, UK. Recommendations for the design and construction of earthworks supporting road pavements are given in the UK Department of Transport's publication *Specification for Road and Bridge Works*, issued by HMSO (London, 1976).

pavement tester A machine used to test aircraft runway pavements by the application of a dynamic load that provides data with which calculations can be made to determine the *modulus of subgrade reaction (k)* and the *flexural rigidity* of the pavement.

pay line The extent of a tunnel, or other underground excavation, beyond which a contractor is not paid for excavation. Its significance lies in the need to control *overbreak*, in that a contractor may be required to replace, at his own expense, rock which has broken away beyond the payline, whether as a result of geological structure or faulty excavation technique.

peak-to-peak strain Peak-to-peak strain is defined in relation to *liquefaction* as the difference between the maximum and the minimum normal or shear *strain* during a given cycle under cyclic loading conditions.
   Reference: 'Definition of terms related to liquefaction', *ASCE Jl Geotech. Eng. Div.*, **104**, Pt GT9, September 1978.

peak pore-pressure ratio A ratio defined in relation to *liquefaction* as the maximum pore-pressure ratio measured during a particular loading sequence.
   Reference: 'Definition of terms related to liquefaction', *ASCE Jl Geotech. Eng. Div.*, **104**, Pt GT9, September 1978.

peak and residual shear strength See *residual and peak shear strength*.

peak strain Peak strain is defined in relation to *liquefaction* as the maximum or minimum *strain* (from the origin or initial state) produced during a particular loading cycle. In cyclically loaded triaxial tests the peak compressional axial (normal) strain is usually reported; however, if the peak extensional axial strain is larger than the peak compressional axial strain, both axial strains should be reported. The peak strain thus includes accumulated extensional or compressional strains accruing during a test.

Reference: 'Definition of terms related to liquefaction', *ASCE Jl Geotech. Eng. Div.*, **104**, Pt GT9, September 1978.

peat Soil comprising dark fibrous or spongy-textured vegetable matter formed from the decomposition of plants. Peat is combustible and is characterised by a high degree of *compressibility*.

pedology The study of the effects of soil by meteorological, climatic and biological factors. It comprises the study of the soil formations resulting from various factors such as the effects of climate, plant and animal life, and topography on the parent constituents; their classification into recognisable soil profiles resulting from the above factors; and the mapping of areas in accordance with this classification.

pellicular water Thin films of water adhering to the surfaces of soil and rock particles at the junctions of interstices in the *zone of aeration, or unsaturated zone* above the capillary fringe.

penetration of cofferdams and caissons The total depth below external ground level reached by a *caisson* or the sheet piles of a *cofferdam*.

penetrometer (apparatus) A defined term for an apparatus consisting of a series of cylindrical rods with a terminating body, called the *penetrometer tip*, and the measuring devices for the determination of the cone resistance, the local side friction and/or the total resistance. Types of penetrometer in use include the following:

(1) Mechanical penetrometer tip: *Dutch mantle cone; Dutch friction sleeve; USSR mantle cone; simple cone; Andina cone; Andina friction sleeve.*

(2) Electric cone penetrometer tip: *Delft electric cone; Delft friction sleeve; Degebo friction sleeve.*

(3) Hydraulic penetrometer tip: *Parez hydraulic cone; Parez friction sleeve.*

Reference: *Proc. 9th Int. Conf. on Soil Mech. and Found. Eng.*, **3**, Appendix A, pp. 99–109, Tokyo, 1977.

peptised bentonite Standard bentonite that has been processed and chemically treated to improve its quality.

percentage air voids, or air content The percentage ratio of the volume of air to total volume of soil.

perched aquifer A body of unconfined *groundwater* separated from the main groundwater by a relatively impermeable stratum. A common feature where alluvial sands and gravels overlie a clay stratum which is, in turn, underlain by permeable strata.

perched water table *See groundwater (perched).*

percussion drilling Also called *cable tool drilling* or *churn drilling*, wherein the hole is advanced by the alternate lifting and dropping of heavy drilling tools to break up the soil or rock and form into a slurry with the *groundwater* or water introduced for this purpose. The slurry is then removed periodically by use of a *bailer, or shell* and a sand pump.

percussive clay cutter   *See shell and auger boring.*

pergelisol   A synonym for *permafrost.*

permafrost   A subsurface zone that is permanently frozen. Permafrost (also called *pergelisol*) is found in high-latitude and high-altitude regions, and may be continuous, discontinuous or sporadic in its areal extent. The surface of the permafrost, or permafrost table, lies a short distance below the ground surface, and above this is located the active layer. The depth to the base of the permafrost is determined by the geothermal heat flow, and may be several thousand metres in some areas—for example, where large thicknesses of sediment have accumulated in the Arctic, such as the major river deltas.

permanent works   A term defined in the British ICE Conditions of Contract as the permanent works to be constructed, completed and maintained in accordance with the contract.

permeability   The property of a porous material which permits a fluid or gas to pass through it when subjected to pressure, such as a hydrostatic force. Where the flow of *groundwater* through soil or rock is laminar (e.g. during conditions of steady seepage), the discharge velocity, v, is equal to ki *(Darcy's law)*, where k = *coefficient of permeability, or hydraulic conductivity* and i = hydraulic gradient. The discharge velocity is the rate of flow across a unit area of a section measured at right angles to the direction of flow. This velocity is a measure of the mass movement of water through the ground and not the actual velocity through the voids. As i is dimensionless, k has the units of velocity, normally f/s or m/s. The coefficient of permeability can be measured in the laboratory by use of constant-head permeameters for the more permeable soils and by falling-head permeameters for fine-grained soils. Such tests should be carried out on samples aligned vertically and horizontally to the *in situ* stratification to measure the anisotropy of the material. However, a more accurate assessment of the mass permeability of the ground will always be obtained by tests carried out in boreholes, the degree of accuracy of which increases from simple falling or rising head tests in single boreholes using, for example, data based on Hvorslev's formulas *(see Figure B.1)*, to full-scale pumping tests from wells with the measurement of groundwater drawdown in observation holes sunk along lines radiating outwards from the centre of the test well. Continuous pumping tests at either one or several pumping rates provide information for the estimation of field permeability, transmissibility and yield.

Values of permeability coefficient may be categorised as follows:

k ($\mu m/s$)*	Degree of permeability	Approximate soil type
>1000	high	clean gravels
10–1000	medium	clean sand and gravel mixtures
0.1–10	low	very fine sands
0.001–0.1	very low	silt, sand and clay mixtures
<0.001	practically impervious	clays

* $1 = \mu m = 10^{-6}$ m

Allen Hazen suggested k for soils of medium permeability to be approximately equal to $100D_{10}^2$ cm/s (D_{10} in cm units). A. Casagrande's unpublished equation $k = 1.4k_{0.85}e^2$ related coefficient of permeability to *void ratio*, where k is the coefficient of permeability of a soil at any given void ratio; $k_{0.85}$ is the coefficient of permeability of the same soil at a void ratio of 0.85; and e is the void ratio.

See also field permeability tests.

References: 'Sub-surface exploration and sampling of soils for civil engineering purposes', M. J. Hvorslev, US Army Corps of Engineers, Waterways Experiment Station, Vicksburg, Mississippi, 1949; 'Physical properties of sands and gravels with reference to their use in filtration', A. Hazen, Report Mass. State Board of Health, 1892.

permeability coefficient The permeability coefficient is defined by Meinzer as the rate of flow in gallons day^{-1} ft^{-2} of cross-section, under unit hydraulic gradient at a temperature of 60° F.

permeability, intrinsic A measure of the relative ease of movement of a fluid under pressure through a porous medium dependent only on the shape and size of the latter's pore spaces.

permeable A condition in which a fluid can flow through a material owing to its porosity or perviousness.

permeable synthetic fabric membranes Woven, non-woven or knitted fabrics generally known as *geotextiles* made from synthetic fibres of polyamide (nylon), polypropylene, polyester (Terylene) and polyethylene. They are used in civil engineering works as drainage blankets and as a means of improving the stability and load-bearing quality of soft soils. A number of proprietary types are available having wide range of physical properties, including Mirafi, Typor, Polyfelt, Fibertex, Terram, Lotrak and Netlon.

permissible stress method A method of defining the factor of safety in design wherein it is applied to the strength of the material and used, for example, in slope stability problems. *See also* **load factor method** and **partial factor method.**

persistence of discontinuity One of the ten parameters selected to describe discontinuities in rock masses, being the discontinuity trace length as observed in an exposure which may give a crude measure of the areal extent or penetration length of a discontinuity. Termination in solid rock or against other discontinuities reduces the persistence.

Reference: International Society for Rock Mechanics. Commission on Standardisation of Laboratory and Field tests, 1977.

pervious Able to transmit water. The use of the term 'pervious', as opposed to 'permeable', usually implies that transmission is via fractures rather than through an interconnected system of pores.

petrofabric The small-scale fabric of a rock, notably the size, shape and orientation of the mineral or sedimentary grains.

petrography The systematic description of rock material by means of its mineralogy, petrofabric, etc.

petrophysics The measurement of physical properties of rocks such as porosity and permeability, and the correlation of these parameters with electrical and sonic characteristics as measured in borehole logs. The study of such features is motivated by the need (particularly in oil geology) to relate borehole records to subsurface reservoir characteristics.

phenomenological model A three-dimensional scale model used to study the physical behaviour of, for example, the stability of a rock slope. The model can be prepared from wood or plaster, saw-cuts being made to simulate the *joint* system.

phi (ϕ) scale A logarithmic scale of particle size used in sedimentology. The size of a grain in phi units is defined as

$$\phi = -\log_2 D$$

where D is the diameter in mm. On this scale 1 mm is 0ϕ and 0.001 mm is 10ϕ, with grains larger than 1 mm having negative values. Boundaries between size classes are gravel–sand at -1ϕ (2 mm), sand–silt at 4ϕ (approximately 0.06 mm) and silt–clay at 9ϕ (approximately 0.002 mm). Intermediate sizes may be calculated either directly or from the fact that each unit increase in ϕ represents a halving of the grain size. The principal advantage of the scale is that since it is logarithmic, a plot of weight frequency against ϕ will be a nearly normal distribution for many sediments, and thus statistical measures such as mean, standard deviation and skewness may be obtained directly in ϕ units.

photogeology Geological information about lithology and structure may be gained from a study of aerial photographs. Photogeology is the application of these principles to produce a regional geological model by integrating photographic evidence of geomorphology, visible rock structure, surface drainage and surface texture. It is normal to use photogeological techniques in conjunction with a limited ground survey, and the method has found greatest application in reconnaissance surveys at a small scale in previously unsurveyed areas.

phreatic surface   The level of the water surface in an unconfined aquifer—i.e. the groundwater table.

phreatic zone   *See zone of saturation.*

pH value   A logarithmic scale used to indicate the relative concentrations of hydrogen ions and hydroxyl ions in an electrolyte.

piezometer   An open or closed tube or other device installed below ground level to measure the hydrostatic or excess water pressure in the pores of saturated soils. In soils with a high *permeability* (e.g. *sands* and *gravels*) an open standpipe (*standpipe-piezometer*) may be used, but for soils of low to medium permeability the time lag is too large, and pneumatic, hydraulic or electrical piezometers are used. *Pneumatic-type piezometers* use a sealed porous tip containing a gas- or fluid-operated valve or diaphragm connected to one or two lines extending to ground level. The valve or diaphragm opens when the pressure applied to the connecting line is equal to the fluid pressure at the tip, which is assumed to be the same as the pore-water pressure. *Hydraulic piezometers*, such as the Casagrande type, comprise a porous tip connected directly to a gauge via a pressure line. Some types are provided with two lines to allow water to be circulated and to de-air the system. *Electrical piezometers* measure the deflection of a diaphragm by an electrical transducer and have the advantage of a very short time lag. They can thus measure rapid changes of pore pressure caused, for example, by earthquakes or pile driving.

piezometric surface   The notional surface formed from the pressure heads in a confined aquifer (i.e. the surface of zero pressure in the aquifer) which lies above the upper boundary.

pile(s)   *See pile foundations.*

pile cap   A concrete block cast on the head of a pile or a group of piles to transmit the load from the structure to the pile or group of piles. Reference: CP 2004 : 1972.

pile-driving analyser   A technique developed at Case Western Reserve University, USA, with the Ohio Highway Administration, for evaluating the bearing capacity of piles and for integrity control. The method allows of the analysis of force and velocity data collected during driving, via strain gauges fixed to the pile. An oscilloscope allows of a visual study of the data received, which can also be recorded on tape for further study.

170

pile foundations   Piles are structural members used to transfer loads through water or ground of low strength to a competent bearing stratum. Piles may be either driven (displacement) or bored (non-displacement), and either preformed (steel tube or section, wood, reinforced or prestressed concrete) or cast-in-place (in situ concrete). They may be 'end bearing piles' (i.e. deriving their carrying capacity from resistance of the point embedded in strong soil or rock) or 'friction piles' (i.e. deriving their carrying capacity from adhesion between the pile shaft and the surrounding ground) or a combination of these factors. A resumé of various methods used is given in *Piling Techniques*, issued by Building and Contract Journals Ltd, London, which is updated from time to time. *See also: alpha pile; Balkan piling system; Benoto piling system; bored and cast-in-situ concrete piles; British Steel Piling Co Ltd cased pile; Concore/Cemcore auger-injected piles; delta pile; Dowsett Prepakt piles; Franki piling systems; Fundex pile; GKN driven pile; Hercules piling system; Hochstrasser-Weise; large-diameter bored pile; Prestcore piles; Raymond piling systems; Simplex pile; sheet pile; steel piles; Tubex pile; under-reamed piles; vibro piles; Western button-bottom pile; West's shell piling system.*

pile frame   A movable steel or timber structure for driving piles on the correct position and alignment by means of a hammer operating in the *leaders* of the frame.
   Reference: CP 2004 : 1972.

pile freeze   *See soil set-up.*

piling hammer   A *single-acting piling hammer* is a hammer raised by steam, compressed air or internal combustion and allowed to fall under gravity. A *double-acting piling hammer* is a hammer operated by steam, compressed air or internal combustion, the energy of its blows being derived mainly from the source of motive power and not from gravity. The most important influence on hammer efficiency is its weight. Although the kinetic energy ($wv^2/2g$) of two hammers may be identical, a heavier hammer travelling with a lower velocity is more efficient than a faster, lighter one. Also, the faster hammer is likely to be noisier and to cause damage to pile heads. Piling hammers can be categorised mainly as air, steam or diesel. Air hammers, either single- or double-acting, are available in a wide range of weights; small hammers are suitable for driving sheet piles, while many large hammers are used mainly for offshore projects such as oil production platform piling. Steam hammers, including single- and double-acting, are mainly only economic when used on large sites. Diesel hammers, basically drop hammers, are made in a variety of sizes for most piling requirements except for large offshore work.

piling helmet   A temporary steel cap placed on top of a precast

concrete pile to minimise damage to the head during driving.
Reference: CP 2004 : 1972.

pillar and stall *See subsidence.*

pinchers A timbering term for a pair of poling boards strutted apart to support a trench wall in firm ground.

pingers Instruments used in marine geophysical surveying to provide a seismic energy source comprising either piezoelectrical crystals or magnetostrictive scrolls which are physically deformed by applying an electric charge or a magnetic field, respectively, which generates a pressure wave. They are used extensively for high-resolution work for resolving near-surface layering. *See also continuous seismic reflection profiling.*

pingo Large frost mound or ice laccolith (up to 300 m diameter and 80 m high) formed by growth of ice due to injection of water entering the soil mantle. Two basic systems exist: (1) open system, prevalent in areas of discontinuous permafrost; and (2) closed system in permafrost areas due to freezing of lakes. In periods of thaw pingos collapse, leaving doughnut-shaped mounds with central marshy areas. Collapsed pingo remnants have been identified in areas of previous periglaciation in the UK.

pipe jacking *See thrust boring, or pipe jacking.*

pipe pile A pile consisting of a length of steel pipe, driven either open-ended or with a shoe.
Reference: CP 2004 : 1972.

pipe step taper piles *See Raymond piling systems.*

pipette analysis A method of particle size analysis applicable to the silt and clay sizes. A sample of soil in this size range is allowed to settle through a water column, and a known volume of the suspension is withdrawn from a fixed depth at known times via a pipette. The particle size distribution of the soil may then be calculated from the weight of material still in suspension after a series of times. The method employs the same principle as the hydrometer method, but is considered to be more accurate, though more time-consuming.

piping The movement of a stream of water and soil below or through a water-retaining structure. This can be caused either by excess hydrostatic pressure being sufficient to produce a *critical hydraulic gradient* in cohesionless soil, or by scour or subsurface erosion that—for example, in a dam—can start at a spring near the downstream toe and proceed upwards along the base of the structure or along a bedding plane, failure occurring as soon as the upstream end of the eroded hole approaches the bottom of the reservoir.

piston corers Piston corers are similar to drop samplers but have an internal piston that remains stationary at or near bed level as the core barrel penetrates. On retrieval the partial vacuum above the sample created by the piston aids its retention. *See drop sampler or corer.*

piston samplers   Samplers used for the recovery of soft cohesive soils and silty fine sands of either floating or fixed type, the piston remaining in contact with the top of the sample to aid its retention as the sampler is withdrawn from the ground. The fixed type has the advantage of being able to be pushed down through soft soils until the required depth of sampling is reached, which obviates the need for a borehole.

pitcher sampler   Basically a *Denison sampler* wherein the inner barrel is spring-loaded so as to automatically keep the cutting edge of the barrel ahead of the coring bit.

pitching   The action of lifting piles and runners and aligning them ready for driving into the ground.

pitting   Non-uniform corrosion formed in the surface of a metal.

plane of décollement   *See gravitational gliding.*

plane strain   A pattern of strain in which all deformation occurs in one coordinate plane, with no strain in the third direction. The assumption of plane strain allows stress and displacement calculations to be simplified. The assumption is normally considered justified for geometries which are elongate in one direction, which is taken to be the third axis. Hence, strains around tunnels are often considered only in the rock face perpendicular to the tunnel, and strains in embankment dams are considered only in cross-sections perpendicular to the run of the dam.

plane stress   A pattern of stress in which all stresses are assumed to act in a plane with zero stress in the third coordinate direction. This condition is met with in slabs and plates in which one dimension is small in comparison with the other two. The small dimension is taken as the axis of zero stress.

plasticity   A material is said to exhibit plasticity if, at a certain level of stress, it undergoes deformation that is irrecoverable when the stress is removed. The magnitude of the deformation is either unrelated to further stress increments (perfect plasticity) or related via a non-linear stress strain law, and so knowledge of the strain distribution in a plastically deformed material does not enable stresses to be calculated. In geotechnics soil can be considered to show plastic behaviour during compression along the virgin line and during shear: this approach is applied in some aspects of critical state theory.

plasticity chart   A chart showing the relationship between *plasticity index (PI)* and *liquid limit* on which the *A-line* is given. *See soil description and soil classification.*

plasticity index (PI)   The numerical difference between the *liquid*

173

limit and the *plastic limit* of a soil. *See Atterberg limits and soil consistency.*

plastic limit (PL) The moisture content at which a soil becomes too dry to be in a plastic condition, as determined by the plastic limit test. *See Atterberg limits and soil consistency.*
Reference: BS 1377 : 1975.

plastic strain That part of the total strain which is irrecoverable on removal of the applied stress.

plate-bearing test A vertical loading test for determining the strength and deformation characteristics of soil, wherein square or circular plates are loaded until failure of the soil occurs or until a specified settlement is reached. Tests carried out in pits or boreholes are particularly useful where soil samples recovered for laboratory analysis would be liable to appreciable sampling disturbance. The loading reaction is normally provided by kentledge or loaded lorry with the settlement being recorded by gauges, located at the quarter points of the plate, suspended from a frame unaffected by the loading action. *See also constant rate of penetration test; incremental-loading plate-bearing test; modulus of subgrade reaction.*

plate, lithospheric A large-scale area of the Earth's surface whose margins are the sites of energetic geological processes and whose interior is quiescent. Plates consist of rigid slabs of either oceanic or continental crust (or both), together with an underlying body of peridotite upper mantle, all of which move as one unit. The total thickness of the plate is about 100 km, compared with an area of tens or hundreds of thousands of square kilometres. Thus, the plates are a very thin skin on the surface of the globe. *See also earthquakes; plate tectonics.*

plate tectonics The theory that the surface of the Earth consists of about a dozen discrete interlocking tectonic 'plates', about 100 km or so in thickness, floating on the Earth's semi-molten mantle. The build-up of friction along the plate boundaries and alternating locking of the plates and release by fracturing of the rock give rise to *earthquakes*. Movement of the plates provides a comprehensive explanation for continental drift, mountain building and volcanism.

playa *See sabkha.*

pneumatic-type piezometers *See piezometer.*

pneumatic-tyred roller A machine used for compaction of soils ranging from clays to granular fills, comprising a ballast box mounted on two axles and provided with an odd number of tyres ranging from 7 to 19. The back wheels are placed out of line with those at the front to allow of maximum coverage of the ground and reduce rutting effects, and are designed to give a kneading action to the fill. Both self-propelled and towed machines are available and roller weights can be varied by the addition of **kentledge**.

pocket penetrometer A hand-held instrument for determining the shear strength of cohesive soil, comprising a short steel rod which is pushed into the soil against the reaction of a calibrated spring.

pockmarks Shallow cone-shaped depressions, normally without a rim, which occur in soft unconsolidated deposits at the sea-bed, thought to be caused by dispersion of localised gas pockets. They are common in the northern North Sea in areas of late glacial and Holocene clay, and are of importance in laying subsea pipelines and siting offshore structures.

podsolic soil A soil from which the iron and aluminium oxides have been leached from its upper horizon and deposited at some greater depth.

point load index/test A field test using portable equipment for determination of rock strength on samples of unprepared core by loading to failure across a diameter at two points. The strength is given in terms of the *point load strength index*, obtained by dividing the force at failure by the square of the core's diameter and corrected, if necessary, for shape and size effects. *See also rock strength index log.*

point load strength index *See point load index/test.*

Poisson's ratio The ratio, v, between the radial strain and the axial strain in an elastic material that is formed into a cylinder and loaded uniaxially. More generally, it is the ratio between the strain in one coordinate direction (due to a stress in that direction) and the strain caused in the other coordinate directions by the same stress. *See stress–strain relationships.*

Poisson solid An isotropic elastic material wherein the Lamé elastic constants λ and μ are equal and having a ***Poisson's ratio*** $v = \frac{1}{4}$. *See stress–strain relationships.*

polar compounds Substances with asymmetrical molecules ranging from water-attracting to water-repellent material—e.g. resin derivatives.

polar projection A map projection such that the pole of the globe lies at the centre of the projection.

pole diagram A projection, usually stereographic, on which are shown the normals to geological surfaces such as beds or joints. The intersection of the normal with the projection sphere is a point, termed a pole, and it is these that are plotted on the diagram. The advantage of pole diagrams lies in the ability to display the orientations of a large number of surfaces, and thus to reveal patterns of clustering, etc. The density of points may be contoured statistically to reveal the significance of the clusters. *See equal area net, or Lambert net.*

poling back The operation of excavating behind timber supports already in position and timbering the new face.
Reference: BS 6031 : 1981.

poling board A flat member in contact with the ground and supporting the face or sides of an excavation, and usually 1 m–1.5 m long.
Reference: CP 2004 : 1972.

pollution A general environmental term for the impairment in quality of air, water, food, noise, etc., which affects our various senses.

polydrill A low-ground-pressure soils investigation rig marketed by Borros AB of Sweden for snow, sand, marshland or undulating terrain, which can perform various drilling and sampling functions.

Poncelet construction A graphical method for the solution of Coulomb's earth pressure theory which does not require trial and error methods.
Reference: 'Mém. sur la stabilité des revêtements et de leurs foundations', V. Poncelet, *Mém. de l'officier du génie*, **13**, 1840.

pore pressure The induced pressure in the pores of a material (e.g. soil) generated by a change in loading. *See effective stress; pore-pressure coefficients A and B; total stress.*

pore-pressure coefficients A and B The properties of a soil which determine the change in pore pressure generated by a change in loading. If the soil cylinder in *Figure P.1*, subjected to triaxial loading, has the vertical stress (σ_1) and horizontal radial stress (σ_3) increased by $\Delta\sigma_1$ and $\Delta\sigma_3$, respectively, an increase in pore pressure from μ to $\mu + \Delta\mu$ will be generated. The pore pressure change ($\Delta\mu$) is the sum of $\bar{A}$ times the change in principal stress ($\Delta\sigma_1 - \Delta\sigma_3$) and B times the change in isotropic stress ($\Delta\sigma_3$)—i.e.

$$\Delta\mu = B\,\Delta\sigma_3 + \bar{A}(\Delta\sigma_1 - \Delta\sigma_3)$$

$\bar{A} = AB$, where A and B are the pore pressure parameters given by Skempton (1954) in the equation

$$\Delta\mu = B[\Delta\sigma_3 + A(\Delta\sigma_1 - \Delta\sigma_3)]$$

When soil is fully saturated, $B = 1$ and $\bar{A} = A$. When soil is unsaturated (dry), $B = 0$.

$$\text{Coefficient } \bar{B} = \frac{\Delta\mu}{\Delta\sigma_1} = B\left[1 - (1 - A)\left(1 - \frac{\Delta\sigma_3}{\Delta\sigma_1}\right)\right]$$

This parameter can be measured directly in the laboratory and is useful in slope stability calculations involving rapid drawdown of the water level.
Reference: 'The pore pressure coefficients A and B', A. W. Skempton, *Géotechnique*, **4**, pp. 143–147, 1954.

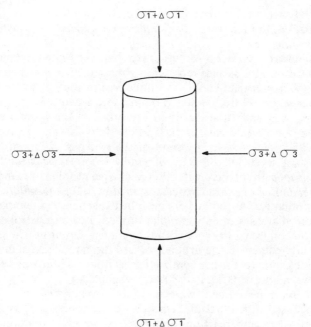

$\sigma_1 + \Delta \sigma_1$

$\sigma_3 + \Delta \sigma_3$ ⟶   ⟵ $\sigma_3 + \Delta \sigma_3$

$\sigma_1 + \Delta \sigma_1$

Figure P.1   Pore pressure coefficients A and B

pore-pressure ratio (r_u)   (1) The ratio between the *in situ* pore pressure (μ) and the total overburden pressure (γz)

$$r_u = \frac{\mu}{\gamma z}$$

(2) Defined in relation to *liquefaction* as the ratio, expressed as a percentage, between the change of pore pressure, $\Delta \mu$, and either the initial effective minor principal stress, $\bar{\sigma}_{3c}$, at the end of primary consolidation (normally used in cyclic triaxial compression tests) or the vertical effective overburden pressure, $\bar{\sigma}_v$ (normally used in simple shear tests and in field studies).

Reference: 'Definition of terms related to liquefaction', *ASCE Jl Geotech. Eng. Div.*, **104**, Pt GT9, September 1978.

porosity   The ratio between the volume of voids in a material and its total volume:

$$\text{porosity } (n) = \frac{e}{1+e}$$

where e is the *void ratio*.

post-hole auger   *See hand auger.*

177

potential energy   *See total energy of a fluid.*

potential infiltration rate   *See infiltration capacity, or potential infiltration rate.*

potentiometric surface   A term synonymous with *piezometric surface*—i.e. the surface representing the static head of an *aquifer*.

pouring-in permeability tests   A simple field method of determining *permeability*, whereby water is poured into a *borehole* to raise the water level and either readings are taken of the water level at various time intervals as it is allowed to fall and equilibrate *(falling-head permeability test)* or the water is maintained at a constant level and the rate of inflow recorded *(constant head permeability test)*. Permeability is calculated in a similar manner as for *bailing-out permeability tests*. *See also field permeability tests.*

power rammer   A pedestrian-controlled machine for compacting confined areas of mainly granular fill, activated by a petrol-driven piston connected to the foot of the rammer. On firing, the piston imparts energy to the ground, causes the rammer to jump and thereby imparts further compaction as it lands. Machine weights between about 100 kg and 800 kg available.

pozzalans   Fine-grained, water-insoluble, chemically active materials used with Portland cement to make concrete. They retard alkali aggregate reactions, reduce heat generation during the hydration of concrete, increase the tensile strength of concrete and, being cheap, can reduce the cost of concrete. Natural pozzalans include clay minerals, zeolites, hydrated oxides of aluminium, pumicite, opal and volcanic glass, tuffs and ashes, while pulverised fuel ash (PFA), an artificial pozzalan, is frequently used in civil engineering works.

Prandtl plastic equilibrium theory   A solution proposed by L. Prandtl in 1920 for the bearing capacity of long footings using plastic equilibrium theory. Originally proposed for metals but shown to be reasonably correct for soils, the hypothesis states that in a soil subjected to loading by a long footing, three distinct zones may be recognised after failure has occurred. In *Figure P.2*, Zone I is an unsheared zone similar to the top of a cylindrical compression test specimen, Zone II is completely sheared and plastic, and Zone III is forced upwards and outwards as a complete unit, owing to passive pressure. Assuming that the shear strength $S = c + \sigma \tan \phi$ and that c is constant, the ultimate bearing capacity of the soil,

$$q_u = \left(\frac{c}{\tan \phi} + \tfrac{1}{2}\gamma b \sqrt{K_p} \right)\left(K_p \varepsilon^{\pi \tan \phi} - 1 \right)$$

where
$$K_p = \frac{1 + \sin \phi}{1 - \sin \phi}$$

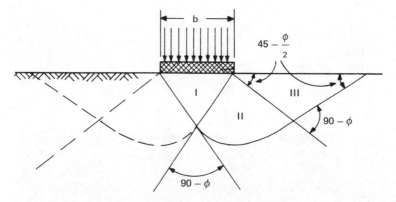

Figure P.2 Prandtl's plastic-equilibrium theory

and the term $\frac{1}{2}\gamma b\sqrt{K_p}$ allows for the increase in strength due to overburden pressure.

Reference: 'Härte Plashecher Körper', L. Prandtl, *Nach. Ges. Wiss. Göttingen*, 1920.

precast pile A reinforced or prestressed concrete pile cast before driving.

Reference: CP 2004 : 1972.

preconsolidation pressure The maximum effective pressure which a soil has been subjected to in its history.

preliminary pile A pile installed before the commencement of the main piling works for the purpose of establishing the suitability of the chosen type of pile and for confirming the design, dimensions and bearing capacity.

Reference: CP 2004 : 1972.

pressure anemometer *See anemometers.*

pressure bulbs A method of displaying stress distributions in the form of contours of equal stresses.

pressure energy *See total energy of a fluid.*

pressure head The head or energy of a particle in a fluid due to the weight of fluid above it. *See also total energy of a fluid.*

pressuremeter test A test used to determine strength and deformation characteristics of soils and rocks by inserting a probe into a borehole to the required depth and expanding it laterally, the required information being obtained from the applied pressures and resulting deformations. *See borehole jack; Camkometer; Menard pressuremeter.*

pressure-operated tide gauge *See automatic tide gauge.*

Prestcore piles Prestcore piles include standard and long-element piles formed by initially sinking a hole by conventional methods

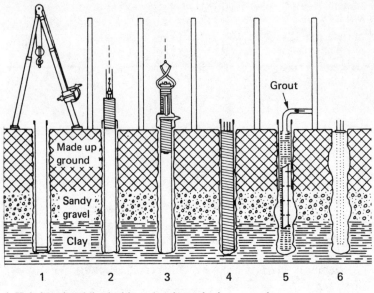

1 Hole bored and lined with steel casing as boring proceeds
2 Assembly of short cylindrical precast concrete units reinforced with spiral binding
3 Building up length of core pile
4 Precast core and vertical reinforcement in final position
5 Steel casing is freed and then withdrawn during grouting process
6 Completed pile

Figure P.3 Stages in the construction of a Prestcore pile

which is supported by a temporary steel casing *(Figure P.3)*. The core pile is formed from cylindrical precast concrete core units which are assembled onto a central steel tube or steel rope and lowered into the lined hole. Reinforcement bars are placed in holes provided in the core units. After placement in the borehole, neat cement grout is fed to the base through a central hole in the core and as it rises in the hole the casing is slowly withdrawn. This allows the grout to grout-in the reinforcement bars, fill the annular space vacated by the casing and also to expel any water from the borehole. The construction provides a pile with a precast core surrounded by dense grout in contact with the surrounding ground. Piles of 350–650 mm nominal diameter are available to carry loads up to 1500 kN. The system is particularly suited to sites where space or headroom is restricted and where difficult subsoil conditions exist. *See also* **pile foundations**.

presumed bearing value   The net loading intensity considered appropriate to the particular type of ground for preliminary design purposes. The particular value is based either on local experience or on calculation from strength tests or field loading tests using a factor of safety against shear failure. Values for various types of ground are given in *Table 1*.
Reference: CP 2004 : 1972.

Prickett analysis   A procedure similar to the *Theis method* for analysing non-steady-state unconfined test data to determine *transmissibility* and *storage coefficients* of an *aquifer*.
Reference: 'Type curve solutions to aquifer tests under water-table conditions', T. A. Prickett, *Groundwater*, **3**, 1965.

Prikkenbeen   A Dutch rock corer comprising an electrically-driven 4.2 tonne sea-bed unit having a core barrel length of 4.2 m, which can sample solid rock in water depths to 60 m.

primary creep   *See creep*.

prime cost item   A contractual term for a sum to be used (wholly or in part) for the execution of the work or for the supply of goods, materials or services for the works and for which the engineer is empowered to instruct the contractor to employ a nominated subcontractor.

priming coat   A medium-grade cut-back bitumen emulsion of milk consistency used in bituminous macadam and road surfacing where sideways shear forces may exist—e.g. bends, superelevations. It promotes adhesion between surface dressing and base course.

principal axis   The direction normal to a principal plane, and thus the direction of the principal stress or strain on that plane.

principal plane   A plane through a point in a stressed (strained) body or medium on which there is no shear stress (strain).

principal strain   At an interior point in a strained body, an arbitrary plane will experience both normal and shear strains. It is possible to find three orthogonal planes of principal strain in the same way as can be found planes of *principal stress* and for the same mathematical reason. Thus, a principal strain is the strain in the direction of the normal to a plane of principal strain.

principal stresses   At an interior point in a stressed body, any general plane will experience both normal and shear stresses. It is, however, possible to find three mutually orthogonal planes, termed *principal planes*, on which the shear stress is zero. The normal stresses on those planes are termed principal stresses, and the directions normal to the planes are termed *axes of principal stress*, each axis being the direction of the principal stress acting on that plane. Mathematically, the relation between the principal stresses and the stresses in the other coordinate directions is as follows. The general stress matrix is a 3×3 symmetric matrix which thus has three real eigenvalues whose eigenvectors form an orthogonal basis for the stress space. Thus, the stresses may be written as a diagonal matrix (the principal stress matrix) which is obtained from the original matrix via a transition matrix whose columns are the eigenvectors (i.e. the principal axes). The principal stress matrix is thus simply the Jordan canonical form of any of the possible general stress matrices.

prismatic compass *See compass.*

Proctor compaction tests *See modified Proctor compaction test, standard Proctor compaction test.*

Proctor curve The relationship between dry density and moisture content obtained when a compaction test is carried out on a cohesive soil *(see Figure A.3).*

Proctor penetration needle A site apparatus used to obtain the approximate difference between the laboratory-defined optimum moisture content of a soil and the actual moisture content of the material after compaction of the material in the field.

Reference: Designation E-22, US Department of the Interior, Bureau of Reclamation, *Earth Manual,* 1963.

proof load The load applied to a selected working pile to confirm that it is suitable for the load at the settlement specified. A proof load should not normally exceed 150 per cent of the working load on the pile.

Reference: CP 2004 : 1972.

proton magnetometer An instrument designed to measure the Earth's magnetic field; types are available for use on land or sea or in the air. As a survey instrument it can be used to provide information on regional geological structures, concealed fault systems, intrusive dykes and sills, massive magnetic ore deposits and metal objects such as wrecks, anchors, pipelines and cables.

proton number The number *(Z)* of electrons orbiting the nucleus of a neutral atom of an element or the number of protons in the nucleus.

provisional sum A contractual term for the sum designated for the execution of work or supply of goods, materials, services or contingencies to be used or not at the discretion of the engineer.

pumping-in permeability tests (1) The *falling-head permeability test* may be a simple test similar to a *pouring-in permeability test* but where more permeable ground requires the quicker addition of water to obtain a sufficiently high differential head of water. In this case, calculation of *permeability* will be as for that test. (2) The *constant-head permeability test* may be carried out in a *borehole*, or in a *piezometer* or *standpipe-piezometer* inserted in a borehole where, depending on the quantity of intake required, water may be added by pumping or pouring. The methods of testing and calculation of permeability are given in the USBR *Earth Manual*, 1974 edition. *See field permeability tests.* (3) *packer tests*, or constant-head tests in rock formations are normally carried out in an unlined borehole in rock strata wherein a section of the borehole is isolated by expanding *packers* at each end of the section required for testing, and water under pressure is introduced within this section. A single packer may be used when testing at the bottom of a borehole. Methods of field testing and calculation of permeability are given in the 1951 tentative edition and the 1974 2nd edition of the USBR *Earth Manual*. *See field permeability tests.*

pumping-out permeability tests (1) The *rising-head permeability test* may be similar to a *baling-out permeability test* but where more permeable ground requires the quicker removal of water to obtain a sufficiently high differential head of water. In this case, the method of test and calculation of *permeability* will be as for that test. (2) The *pumping-out permeability test* from the trial pit (constant-head test) is a test primarily for use in a trial pit penetrating below the water table but also used in boreholes where permeability is too high for the *rising-head permeability test* to be practicable. Permeability is affected by the plan shape of the pit and whether or not a local source of water exists such as a canal or river. Solutions for these conditions are described by C. I. Mansur and R. I. Kaufman in Chapter 3 ('Dewatering') of *Foundation Engineering* edited by G. A. Leonards, McGraw-Hill, 1962. (3) The *pumping well with drawdown observation boreholes test* is a constant-head full-scale pumping test wherein water is pumped from a well and measurement is made of the water-table *drawdown* in a number of observation holes set along lines radiating outwards from the pump well. This type of test gives the most accurate assessment of the mass permeability of the ground, solutions for which are also given in *Foundation Engineering*. *See also field permeability tests.*

pumping-out test from trial pit *See field permeability tests; pumping-out permeability tests.*

pumping test In general, a test normally involving the pumping of water from a *well* at a steady known rate and measuring the *drawdown* of the *groundwater* level at known distances from the well, in order to determine the *permeability, transmissibility* and storage capacity of the ground. It is used to evaluate groundwater resources and dewatering requirements. *See also Cooper–Jacob analysis; pumping-in permeability test; pumping-out permeability test; Theis method.*

pumping well with drawdown observation boreholes *See pumping-out permeability tests.*

puncheons Vertical *struts* transmitting the weight of the *bracing* to

the excavated ground surface inside a *cofferdam*.
Reference: CP 2004 : 1972.

push-in pressuremeter (PIP)   *In situ* testing equipment, similar to the *Camkometer*, developed by the British Building Research Establishment for determination of strength and deformation characteristics of soil. It comprises three main parts: the pressuremeter, consisting of a steel cylinder with a cutting shoe and housing an inflatable membrane; the pressure developer, comprising an inflator mechanism and pressure *transducers* to measure the pressures exerted by the membrane against the soil; and the control and data aquisition equipment. The PIP can be pushed into the ground by use of either the McClelland Engineers' *Stingray* system or the Fugro *drillstring anchor* system.

PVC tubing   Tubing made from polyvinylchloride.

P-wave   An elastic body wave in which the motion of the particles is in the direction of propagation. *See also compressional wave; transverse wave.*

p–y method of pile design   A method of determining the deflection of a laterally loaded pile by considering it to be replaced by a large number of independently loaded segments through each of which a single resultant lateral force or resistance, p, is considered to replace the non-linear stress distribution produced around the pile by the deflection, y. For a given depth below ground level, the p–y relationship depends on *soil strength* and density, K_0 and K_a of granular material, and pile diameter.

pycnometer   A device with which the moisture content of granular soils can be rapidly determined.

pyroclastics   Uncemented volcanic dusts.

Q

quartering   The reduction in quantity of a large sample of material by dividing a circular heap, by diameters at right angles, into four approximately equal parts, removing two diagonally opposite quarters, and mixing the two remaining quarters intimately together so as to obtain a truly representative half of the original mass. The process is repeated until a sample of the required size is obtained.
Reference: BS 1377 : 1975.

quasi-static penetration test   *See cone penetration test (CPT).*

quick clay   A sensitive clay. *See sensitivity (S_t).*

quick sands   Very loose saturated sands either disturbed by vibration or subjected to an upward flow of water such that the sand grains are buoyed up and the bearing capacity of the soil is reduced

184

to a very low value. Fine sands with a *coefficient of uniformity* less than about 5 and an *effective grain size* less than 0.1 mm have been found to be most affected by this condition.

Q-wave A *Love wave*.

R

radar (radio detection and ranging) A time-measuring pulsed transmission system using electromagnetic waves of 3×10^9 Hz and 10×10^9 Hz frequency. It is basically intended as navigational aid but can be used for position fixing in marine surveys.

radial drainage The idealised flow pattern that is assumed to develop around a *well* or *borehole*, in which the water flow is towards the centre from all directions. In practice, this pattern probably does not develop, owing to lateral inhomogeneities, but is assumed to do so for ease of calculation.

radiation log A *well* or *borehole* record of natural or induced radiation. *See well logging.*

radioactivity The spontaneous disintegration of unstable atomic nuclei, which is accompanied by emissions of alpha and beta particles and/or gamma rays.

radioactivity log A log made in a *well* or *borehole* wherein measurements are made of natural or induced radiation. *See well logging.*

radioactivity survey The surveying or mapping of the distribution of radioactive elements such as uranium, potassium or thorium by measuring the emission of *gamma rays* with a suitable instrument such as a Geiger counter.

radiocarbon dating *See carbon-14 dating.*

radiotracers Radioisotopes which can be introduced to water to study various water movement problems such as reservoir and canal leakage. The emitted radiation is measured by various instruments such as a Geiger–Müller counter. Popular radiotracers are *bromine-82 (^{82}Br), iodine-131 (^{131}I), gold-198 (^{198}Au)* and *chromium-51 (^{51}Cr)*. Radiotracers have advantages over chemical tracers since they can be introduced in minute quantities owing to their very high sensitivity to detection by appropriate measuring equipment.

radius of influence The distance (R) from a pumping well where discernible *drawdown* of the *water table* occurs:

$$R \simeq 3000 h_0 \sqrt{k}$$

where R = the radius of influence (m); h_0 = drawdown at the pumping well (m); and k = the permeability of the ground (m/s). The shape of the *cone of water-table depression* produced by

185

pumping from a well, as determined by electrical analogy, is given in *Figure D.3*, from which the percentage drawdown at a given distance from the well can be estimated.

raft foundation A foundation continuous in two directions, usually covering an area equal to or greater than the base area of the structure.

Reference: CP 2004 : 1972.

RAFTS A computer program for the analysis of soil–structure interaction problems.

Reference: *Advances in Engineering Software*, Vol. 1, No. 1, 1978.

raking pile A pile installed at an inclination to the vertical.

Reference: CP 2004 : 1972.

raking shore An inclined *strut* or series of struts in the same vertical plane placed against a wall to restrain it from lateral movement. Its upper end abuts against a *needle* and its lower end bears upon a *sole plate*, the whole of the series of struts being braced or laced together.

Reference: CP 2004 : 1972.

RALOG A marine survey instrument for distance measurement usually used in conjunction with *sounding* equipment to give depth–distance data and a true-to-scale sea-bed profile.

ram sounding method A dynamic penetration/sounding test in which a 32 mm diameter 90 degree angle point attached to rods is driven into the ground by a 63.5 kg hammer dropping 50 cm.

Rankine's earth pressure theory The classical *earth pressure* theory presented by Rankine for cohesionless soils wherein expressions were given for the *active pressure* and *passive pressure* generated by soils adjacent to retaining walls. It assumed an interrelationship between vertical and lateral pressures on vertical planes within the soil adjacent to the wall and that the presence of the wall did not cause changes in the shearing stresses at the contact surfaces between the wall and the retained soil. For a homogeneous soil inclined at an angle β behind the wall *(Figure R.1)*, the *active earth pressure* acts at any depth below the soil surface parallel to the surface. For the simple case where the soil is horizontal behind the wall $(\beta = 0)$ the unit active pressure, p, is given by

$$p = \gamma z \frac{1 - \sin \phi}{1 + \sin \phi} = \gamma z \tan^2 \left(45° - \frac{\phi}{2} \right)$$

and the resultant thrust, P, is given by

$$P = \tfrac{1}{2}\gamma H^2 \frac{1 - \sin \phi}{1 + \sin \phi} = \tfrac{1}{2}\gamma H^2 \tan^2 \left(45° - \frac{\phi}{2} \right)$$

where ϕ = the angle of internal friction of the soil; H = the height of

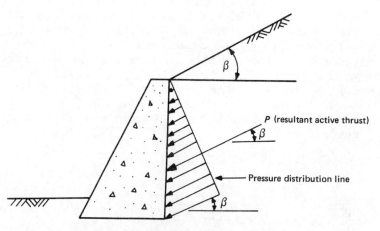

Figure R.1 Rankine earth-pressure theory

the wall; z = the depth below the top of the wall; and γ = the density of the soil. The expression

$$\frac{1-\sin \phi}{1+\sin \phi}$$

is termed the *coefficient of **active earth pressure*** and is denoted by K_a. Similarly, for the passive case where the wall is pushed towards the soil—as occurs in front of the toe of the wall—the unit passive resistance P_p for $\beta = 0$ is given by

$$P_p = \gamma z \frac{1+\sin \phi}{1-\sin \phi} = \gamma z \tan^2 \left(45° + \frac{\phi}{2} \right)$$

and the resultant passive resistance, P_p, is given by

$$P_p = \tfrac{1}{2}\gamma H^2 \frac{1+\sin \phi}{1-\sin \phi} = \tfrac{1}{2}\gamma H^2 \tan^2 \left(45° + \frac{\phi}{2} \right)$$

The expression

$$\frac{1+\sin \phi}{1-\sin \phi}$$

is termed the *coefficient of passive resistance. See also **Coulomb's earth pressure theory**.*

Reference: 'On the stability of loose earth', W. J. M. Rankine, *Phil. Trans. R. Soc.*, London, 1857.

rapid drawdown A term referring to the rate of drawdown in the level of water impounded in a reservoir. The term 'rapid' is relative to the time taken by the phreatic surface in the embankment to follow the falling water level in the reservoir, which may take several weeks or even months, being dependent on the *permeability* of the material forming the embankment.

rate of discharge The volume of water seeping through a given area per unit of time.

rate of secondary consolidation The slope of the final portion of the change of volume per unit volume–time curve, in a semi-logarithmic plot obtained from a *consolidation* test on a soil sample, and equal to:

$$\frac{-\Delta e}{[(1 + e_0)\Delta \log t]}$$

where Δe = the change in the void ratio; e_0 = the initial void ratio; and $\Delta \log t$ = logarithm to base 10 of change in time.
Reference: *Proc. 9th Int. Conf. on Soil Mech. and Found. Eng.*, **3**, p. 164, Tokyo, 1977.

ravelling A type of rock failure that occurs in highly fractured rock where superficial disintegration predominates over any major instability.

Raydist A medium-range electronic positioning system involving the emission of three or more continuous waves differing by an audio frequency, the phase differences of which are used to determine position.

Rayleigh wave A *seismic wave* propagated along a free surface in which the particles have a retrograde elliptical motion—i.e. the particle moves opposite to the direction of propagation at the top of its elliptical path.

Raymond cast-in-place concrete piles *See Raymond piling systems.*

Raymond cylinder piles *See Raymond piling systems.*

Raymond piling systems (1) *Raymond cast-in-place concrete piles* are installed by driving a steel core and shell to the required resistance and/or penetration, withdrawing the core, internally inspecting the shell left in the ground and filling with concrete. This system includes:

(*a*) *Step taper piles*, which employ closed-ended spirally corrugated steel casings to provide strength against ground pressures, ranging in size from 220 mm to 440 mm diameter. The steel core is of similar shape to the casing, which ensures that the maximum driving energy is transmitted throughout the length of the pile.

(*b*) *Pipe-step taper piles*, which can be used where the normal step-taper pile is of insufficient length and are similar, except that the bottom section is replaced by a 273 mm diameter closed-ended steel pipe which can be of any length.

188

(c) *Wood-step taper piles* comprise a step-taper shell cast-in-place concrete upper section and a timber pile lower section. The timber is usually untreated and is driven entirely below the standing groundwater table. The construction combines the permanence of concrete piles with the low cost of untreated timber piles.

(d) Uniform taper piles. These comprise heavily tapered steel shells which are driven by an expansible type mandrel which after driving can be contracted to permit withdrawal without disturbing the driven shell. A 203 mm diameter flat-plate is used at the pile point and a maximum pile length of about 11 m is permissible.

(2) *Raymond cylinder piles.* Circular hollow prestressed concrete cylinder piles of 914 mm and 1370 mm diameter are available. Such piles of considerable length have been driven in marine situations. (3) *Raymond regulated injection piles* are cast-in-place concrete piles for relatively light loadings, formed by drilling with a continuous helical flight auger in stable ground and injecting concrete under pressure through the drill string. (4) Other piling systems including H-beam, pipe, steel sheet and timber are undertaken.

Raymond regulated injection piles *See Raymond piling systems.*

Raymond standard test *See standard penetration test.*

reclaimed aggregate material (RAM) Removed and/or processed pavement material containing no re-usable binding agent.

reclaimed asphalt pavement (RAP) Removed and/or processed pavement materials containing asphalt and aggregate.

recycling The re-use, usually after some processing, of a material that has already served its first-intended purpose. Recycling methods include *hot mix, cold mix* and *surface* methods.

redox potential The stability of an element in a particular state of oxidation depends on the energy change involved in adding or subtracting electrons. A quantitative measure for this energy state is termed the *oxidation–reduction potential* or redox potential. It is defined as that potential necessary in a cell to produce oxidation at the *anode* and reduction at the *cathode*, relative to a *standard hydrogen electrode.* Such an *electrode* comprises hydrogen bubbling over platinum and producing a standard concentration of hydrogen ions.

reduced level The elevation of a point relative to the datum adopted.

reduction factor for piles in groups The reduction factor applied to the carrying capacity of a group of piles due to interaction of stressed zones surrounding individual piles within the supporting media. For essentially end-bearing piles the reduction factor is zero but for essentially friction piles the reduction factor may exceed 25 per cent for large groups.

Reference: 'Experiments with model piles in groups', T. Whitaker, *Géotechnique*, **7**, No. 4, 1957.

reflected sound   The increase in *sound* due to reflected waves; where a site is enclosed by non-absorbent walls on one or more sides, due allowance should be made for an increase in the sound due to wave reflection. A practical approach is to consider the reflected sound attenuated for increased distance travelled with no allowance for absorption; the total sound is unlikely to be greater than this figure. *See also decibel; equivalent continuous sound level; noise; sound attenuation; sound level; total sound level.*

refraction profiling   A *refractor survey* method.

refraction seismic prospecting   A *refractor survey* method. *See also seismic methods of surveying.*

refraction wave   Waves that enter and leave 'high-velocity' strata (*refractors*) at about the *critical angle* and travel roughly parallel to their surface. *See also seismic methods of surveying; head waves; Mintrop wave; critical angle; refractor survey.*

refractors   Layers or strata, in which waves travel at a higher velocity than in overlying materials.

refractor survey   A technique for mapping geological structures by using *head waves*. Measurement of the arrival times of the waves at the ground surface can be interpreted to give the depths of the refractors. *See also seismic methods of surveying.*

regression   The reduction in strength of a soil due to a very gradual increase in its natural moisture content.

reinforced earth   A composite material formed by reinforcing earth (soil) in a manner similar to that in which plain concrete is reinforced with steel bars to give reinforced concrete *(Figure R.2)*. Soil has little or no tensile strength, and reinforcement, usually in the form of metal strips, is generally laid horizontally within the

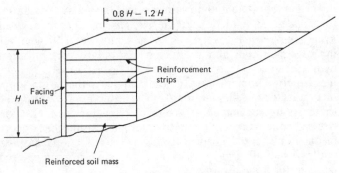

Figure R.2   Principles of reinforced earth

earth mass to resist tension in the matrix, the strips being attached to vertical facing panels. The basic mechanism is the friction between the earth and the strips, and minimum frictional characteristics for soil of 25 degree minimum angle of friction and 15 per cent maximum passing a No. 200 sieve have been quoted. Well over 2000 engineering structures incorporating reinforced earth have been constructed throughout the world.

relative compaction   The percentage ratio between the *dry density of* the *soil* and the *maximum dry density* of that soil, as determined by a specified laboratory compaction test.
   Reference: BS 1377 : 1975.

relative consistency (C_r)   *See Atterberg limits and soil consistency.*

relative density   The relative density of a material is its natural density relative to the loosest and densest possible states of compaction of the material. It is expressed as

$$D_r = \left(\frac{\gamma_{dmax.}}{\gamma_d}\right)\left(\frac{\gamma_d - \gamma_{dmin.}}{\gamma_{dmax.} - \gamma_{dmin.}}\right)$$

where $\gamma_{dmax.}$ = the 'densest' dry unit weight; $\gamma_{dmin.}$ = the 'loosest' dry unit weight; and γ_d = the natural dry unit weight. D_r can also be expressed in terms of the *void ratio* as

$$\frac{e_{max} - e}{e_{max} - e_{min}}$$

where e_{max} = the void ratio of the loosest state of compaction; e_{min} = the void ratio of the densest state of compaction; and e = the natural void ratio. *See also standard penetration test.*

relative humidity   The actual water vapour pressure expressed as a percentage of the saturated water vapour pressure at the same temperature. It is a measure of the dampness of the air and can be measured from wet and dry bulb temperatures or directly using a hygroscopic material.

relief well   A well sunk through impermeable strata to permeable soil or rock to allow of dissipation of potentially high pressures that might develop where overburden is removed—e.g. at the base of an excavation.

remote sensing   The collection of data by systems which are not in direct contact with the objects or phenomena under investigation. The term relates particularly to aerial methods such as aerial photography, TV and infra-red observations using aircraft and satellites. Phases of progress range from around 1860, when there was a slow recognition that photographs taken from the air would be useful for mapping purposes, to modern data-gathering techniques using satellites for weather observation and prediction, for

191

geomorphic–geologic mapping and for determination of agricultural resources, etc. In general, the major objective is to detect and record energy in a selective portion of the *electromagnetic spectrum*, the sensors acquiring imagery by detecting or sensing levels of emitted and/or reflected radiation.

remoulded undrained shear strength (c_r) The shear strength of a remoulded soil in an undrained condition.

repeatability The ability of a technician using the same apparatus at a laboratory to obtain successively similar test results on the same soil.

reporting The systematic recording of factual information obtained in the field and laboratory together with the analysis and interpretation of that information, the conclusions drawn and the recommendations given on which future action may be formulated.

reproducibility A measure of the variation in the results obtained by operators carrying out tests in different laboratories but on the same material.

resection The method of locating the horizontal position of a survey station by the intersection of lines indicating the direction from other stations.

residual angle of internal friction (ϕ'_R) The residual angle of internal friction is a shear strength parameter with respect to effective stresses in the equation:

$$\text{residual shear strength} = C'_R + \sigma' \tan \phi'_R$$

where C'_R = the residual cohesion; and σ' = the effective pressure.

residual cohesion (C_R) The residual cohesion intercept is a shear strength parameter with respect to effective stresses in the equation

$$\text{residual shear strength } \tau_R = C'_R + \sigma' \tan \phi'_R$$

where ϕ'_R = the residual angle of internal fraction; and σ' = the effective pressure.

residual factor (R)

$$R = \frac{S_f - \bar{S}}{S_f - S_r}$$

where S_f = the peak shear strength of the soil; $\bar{S}$ = the average shear strength around the failure surface; and S_r = the residual shear strength. R lies between peak and residual strengths. *See also residual and peak shear strength.*

residual and peak shear strength As a clay soil is sheared, an initial peak shear strength (S_f) is reached after which the clay strain-softens, reaching, after large displacements, a lower residual value (S_r). The latter value corresponds to the resistance to sliding on an

192

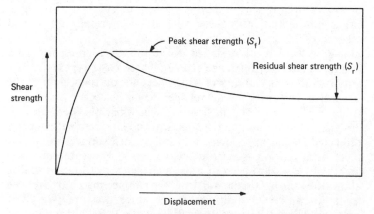

Figure R.3   Residual and peak shear strength

established shear plane—*see Figure R.3*. Dense granular soils exhibit a similar initial peak strength followed by a lower residual value after large strain. *See also* **residual factor (R)**.

residual shear strength (τ_R)   The ultimate shear strength in a rupture plane which a soil maintains at large displacement.

residual soil   Soil formed *in situ* by weathering of the local bedrock.

resin–gypsum cement   A mixture used for grouting where fast or controlled setting is required.

resins   Naturally occurring amorphous organic compounds, insoluble in water but soluble in certain organic solvents such as shellac. They also exist as synthetic plastic substances produced as a result of polymerisation.

resistance   The property of a material by virtue of which it resists the flow of electricity through it. The unit of measurement is the ohm.

resistivity   The *electrical resistivity* of a material is the resistance across opposing faces of a unit cube of the material. It is a property of the material, and does not depend on the geometry of the object which the material composes. In geotechnics the resistivity of soil layers may be measured by surface electrodes and the depths to interfaces determined from changes in resistivity as the electrode spacing is altered. Good resistive contacts are obtained between saturated and dry soils, and between loose soil and hard rock. Thus, the method is of greatest value in finding the depth to the water table or to rockhead. *See Table 10.*

resistivity log   (1) Well or borehole records obtained with sondes using resistivity methods. *See induction log.* (2) Records of the resistivity method obtained from surface investigations.

resonant-column test   A test to study the effects of variations in *stress*

193

or *strain* amplitudes while a cylindrical column of soil is vibrated in either the longitudinal or the torsional mode, normally in a triaxial cell.

retaining wall A wall designed to resist lateral pressure from retained material (Reference: CP 2). Three basic types exist: (1) gravity walls, the stability of which depends on the weight of the structures, with little or no reliance being placed on the tensile strength of the wall; (2) flexible walls in which the stem may be designed as a cantilever and/or a beam; and (3) revetments wherein protected facings are provided on soil or rock surfaces to protect them from erosion—i.e. scour, wave action or weathering. The ability of the wall to bend or move sufficiently by tilting or sliding allows the pressure on the wall from the retained material to reduce from the 'at rest' condition to the active pressure (*see earth pressure*). Types of gravity wall include mass concrete, unit construction—e.g. masonry, brickwork and dry rubble; precast units; and caissons. Flexible walls include reinforced concrete cantilevers, counterforts, buttressed and precast concrete designs; sheet pile design: diaphragm, secant and contiguous pile walls; and horizontal planking of wood or concrete spanning between steel, concrete or wood piles. Types of revetment include grass and other plants either grown *in situ* or placed as turfs; faggots or fascine mattresses of willow and brushwood; tipped stone; pitching with concrete blocks, bagwork, slabs or brickwork; stone *rip-rap;* and special concrete shapes such as tetrapods. Other forms of wall include *reinforced earth* walls and timber or reinforced concrete crib walls, wherein individual units are built up to form box-like structures into which backfill is placed to form an integral part of the retaining structure.

reverse circulation drilling A form of rotary drilling incorporating a reversed flow of drilling fluid to that employed in conventional *rotary drilling*. The fluid is run under gravity from the mud pit to the *borehole*, whence it moves down the annular space around the drill pipe to the bottom of the hole, where it picks up the rock cuttings and enters the drill pipe through ports in the drill bit. It then rises up the drill pipe back to the mud pit via the rig pump. Drilling *mud* and other additives are not normally added to the circulating water, and the stability of the borehole is maintained by keeping the water level at ground level or above so that the hydrostatic pressure of the water column plus the inertia of the drilling fluid as it moves down the borehole supports the sides of the hole. The method is relatively inexpensive for drilling large-diameter holes in soft and unconsolidated deposits, and drilling costs do not increase significantly for increasing diameter of hole. Limiting factors in the use of the system include too high a water table, the need for adequate supply of drilling water, the presence of stiff cohesive deposits and high stone content which may require frequent extraction to permit drilling to continue.

revert An organic polymer produced by the Johnson Division of Universal Oil Products (UOP) from food-grade guar beans which can be added to fresh or saline water to provide a time-dependent viscous drilling fluid as an alternative to native clay or bentonite. The viscosity is maintained during the drilling operation but after about 2–4 days, depending on the water salinity, enzymatic action by the soil bacteria breaks down the gel and the viscosity of the fluid reduces to about that of water. Its use in wells sunk for water supply or construction dewatering removes the problem of caking of the sides of the well when clay suspensions are used as a drilling fluid with the consequent reduction in soil *permeability* and well yield. *Fast-break*, another Johnson additive, can be used to rapidly reduce viscosity of the revert mix.

revised earthquake intensity scale A revision to the *modified Mercalli (MM) Scale* in which intensity is based more on deterministic than purely subjective analysis.

Reference: 'Re-evaluation of modified Mercalli intensity scale for earthquakes using distance as determinant', R. J. Brazee, *Bull. Seismological Soc. Am.*, **69**, No. 3, June 1979.

Reynold's number A dimensionless ratio—symbol (*Re*)—between inertial and viscous forces used for distinguishing *laminar* from *turbulent flow*.

$$Re = \frac{\gamma v D}{u}$$

where $\gamma =$ the density of the fluid; $v =$ the velocity of the fluid; $D =$ the diameter of the pipe; $u =$ viscosity.

rheid The rheid concept is used in structural geology to indicate the expected contribution of long-term viscous flow to the deformation of a stratum. The rock is said to be a rheid if the viscous deformation exceeds other deformation by a factor of 1000. The length of time necessary for this to occur is termed the rheidity of the rock. Rocks which have behaved as rheids usually exhibit an exaggerated style of flow folding termed *rheomorphic folding*.

rheology The study of the stress–strain–time relationships for materials and their classification into groups such as elastic, plastic, viscous, etc. More complex behaviour is often modelled mathematically by considering the behaviour of idealised elements such as springs and dashpots linked into combinations that simulate the observed behaviour of materials. It should be noted that rheology is the study of deformation, and as such does not consider theories of failure.

rheomorphic folding A style of flow folding shown by rocks that have undergone excessive viscous deformation—for example, salt in salt domes, and granite in some intrusions. The style is characterised by drawn-out fold noses, multiple refolded structures and considerable thickness changes along individual layers. In extreme cases balls and other mixing structures may be seen. *See also rheid.*

rheopexy The property of certain soils such as bentonite to produce accelerated gelling when agitated, owing to their platy structure.

Richter scale of magnitude A scale of earthquake magnitude determined by C. F. Richter and shown to be of the form $M = \log_{10} A$ where M is the magnitude and A is the trace amplitude measured in μm for an epicentral distance of 100 km. *See also earthquakes.*

Riedel shears Secondary shear fractures that in a brittle material form at an angle to the primary direction of shear displacement.

riffling The reduction in quantity of a large sample of material by dividing the mass into two approximately equal halves by passing the sample through an appropriately sized riffle (or riffle-box). The process is repeated until a sample of the required size is obtained. *See also quartering.*
Reference: BS 1377 : 1975.

rigid pavements Concrete slabs, either reinforced or unreinforced, used for airfield runways, roads and hard standings. *See also pavement design.*

ring shear aparatus *See Bromhead ring shear.*

ripper A two-wheeled trailer similar to an agricultural tined cultivator but of appreciably more robust construction.

rip-rap (protection for slopes) Broken stone placed on earth surfaces for their protection against the action of water.

rising-head permeability test *See field permeability tests; pumping-out permeability tests.*

road base The layer in a pavement immediately below the surfacing and above the *sub-base*. In flexible pavement design it may comprise lean concrete, soil cement, cement-bound granular material, wet-mix and dry-bound macadam, bitumen macadam, tarmacadam or rolled asphalt. In rigid pavement design a plain or reinforced concrete slab overlies the sub-base and generally performs the functions of both road base and surfacing. *See also pavement design.*

rock bit drilling A *rotary probe drilling* method first used in the oil industry and essentially a *rotary drilling* process using a tricone bit in which toothed wheels run around the full face of the *borehole* removing rock cuttings. Flushing media may be water, air or *mud*, and may be used through all materials. A heavy rig is required to drill through rock, and the method is often used to penetrate obstructions in difficult ground such as slag tips and hard fill materials.

rock bolt   A steel shaft, up to about 5 m in length and a few centimetres diameter, used to secure loose rock on the face of an excavation. The bolt is placed in a drill hole that extends back into sound rock and is secured either mechanically or with epoxy resin or grout. The bolt is then placed in tension by tightening a nut down onto a plate whose function is to spread the load into the surrounding rock. The rock is thus placed in compression and so is secured against failure.

rock burst   In brittle rocks stored strain energy may be released violently during excavation. Such a release is termed a rock burst or bump, and is potentially hazardous. Rock bursts are particularly violent in areas where there is a high *in situ* stress.

rock mass   A large body of rock replete with discontinuities and inhomogeneities. Thus, the rock mass is the rock in its typical field appearance, and its properties in that state are the *mass properties*. These may be more significant in engineering work than the more easily measured *material properties*. There have recently been recommendations on the method of reporting rock masses and their characteristics.

Reference: 'The description of rock masses for engineering purposes', *Qly Jl Eng. Geol.*, **10**, pp. 355–388, 1977.

rock mechanics   The science of rock mechanics comprises several interrelated disciplines. In a strict sense, it is the application of classical continuum mechanics to rock bodies, together with the establishment of failure criteria and the investigation of the effects of fissuring. The experimental basis for field and laboratory studies is also included. In broader terms, the allied discipline of rock engineering, which is the application of rock mechanics to design problems in civil and mining engineering, is also considered a part of the subject. The principles of rock mechanics also find application in structural geology, particularly in those problems that allow of a quantitative evaluation of the mechanical behaviour of the rocks in question.

rock quality   A descriptive term employed to indicate the condition of a rock mass, particularly its state of fissuring. The term is usually combined with a numerical value, the *rock quality designation (RQD)*. *See Figure R.4.*

rock quality designation (RQD)   A parameter used in the description of rock core recovered from a *borehole* and defined as the summed length of the pieces of sound core over 0.1 m long expressed as a

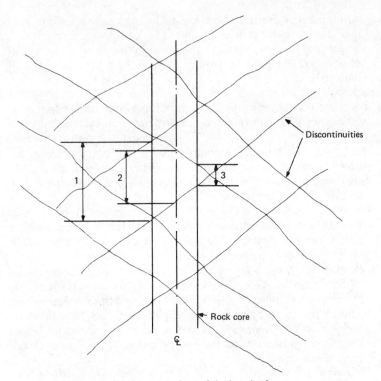

Examples of three possible interpretations of the length of core
pieces:
1 Measured from tip to tip
2 Measured along centre-line of core
3 Fully cylindrical length of core

Figure R.4   Rock quality designation

percentage of the length drilled. The International Society for Rock
Mechanics recommends that the length of individual core pieces
should be assessed along the centre-line of the core, so that any
discontinuity that happens to parallel the drill hole does not unduly
penalise the RQD values of an otherwise massive rock mass
(*Figure R.4*). See Table 9.

Reference: Commission on Standardisation of Laboratory and
Field Tests. Suggested methods for the quantitative description of
discontinuities in rock masses, 1977.

rock-roller bit   A type of drill bit with conical-shaped rolling cutting
edges that rotate.

rock strength   The following scale of rock strength is given in the
British Code of Practice for Site Investigations (BS 5930: 1981):

198

Term	Compressive strength (MN/m²)
Extremely strong	>200
Very strong	100–200
Strong	50–100
Moderately strong	12.5–50
Moderately weak	5–12.5
Weak	1.25–5
Very weak	<1.25

Apart from the addition of the 'extremely strong' term in the Code, the terms are identical with the recommendations given by the Geological Society of London's group working party set up to study 'The description of rock masses for engineering purposes', (*Qly Jl Eng. Geol.*, **10**, pp. 355–388, 1977). *See also* **soil strength;** *Table 7.*

rock strength index log *Point load index test* equipment for determination of *point load strength index* on samples of unprepared core in the field.

roentgen A unit of X-ray or gamma ray dose for which the resulting ionisation under standard conditions of temperature and pressure liberates a charge of approximately 2.58×10^{-4} coulomb per kg of air. The roentgen is a measure of the amount of radiation regardless of the rate at which the quantity is produced, and thus the intensity of radiation must be expressed as roentgens per unit time.

roller bits *See* **rolling-cutter bits.**

roller, pneumatic *See* **pneumatic-tyred roller.**

rolling-cutter bits Also known as *roller bits* or *cone bits*, rolling-cutter bits used for drilling in mainly consolidated strata for rapid open-hole boring where recovery of core is not required. They are available in double- or triple-cone (tricone) types with a variety of tooth configurations to suit various rock types, but fall basically into four main types—viz. very soft, medium to hard, hard and very hard formation bits.

rooter A heavy-duty *ripper* but usually having only two or three tines, used for clearance of obstructions on site.

rotary core drilling A method of drilling used to obtain relatively undisturbed core of the rock material in order to examine the geological succession beneath a site. The *borehole* is advanced by rotating a *core barrel* and allowing the grinding action of an annular *bit* set with diamonds or other hard material to cut into the rock. Core recovery generally improves with the diameter of the core but drilling costs increase sharply. Core obtained for geotechnical purposes should not normally be less than 54 mm diameter (NX size) and preferably 76 mm diameter (HX size) or larger.

Rotary core drilling equipment *(see Table 16)* is commonly described by means of alphabetical prefixes. The components of basic rotary core drilling equipment are specified in BS 4019: Part 1 (1966), and information reproduced from this publication is given in *Table 17.* In this table the first letter in every case (with the exception of XRT) is a size identification symbol. The letter X is used to denote the nesting properties of each size—for example, an NWX core barrel will pass through NX casing and will drill a hole to take BX casing. The letter M denotes a range of core barrels having a design different from that of the X series but having the same nesting properties as the X series. The letter F applied to the larger sizes of core barrels denotes that they are of the face discharge type. The letter W denotes a design of drill rod and core barrel (head thread only) which superseded the X series core barrels (now WX series) and E, A, B, N and H drill rods. The letter T is applied only to the XR size range of equipment (i.e. XRT) and has no general significance.

rotary core samples Samples obtained by *rotary core drilling*.

rotary drilling A method of drilling whereby a rotating *bit* is pushed into the ground or subjected to repetitive hammering. The bit is attached to a hollow-stem pipe through which a drilling fluid—air, water or *mud*—is pumped. The fluid discharges at the bit, thereby cooling and lubricating it, and then circulates back up to the surface between the drill stem and the side of the borehole, carrying with it the soil and rock cuttings. A wide range of rotary drilling rigs have evolved, the individual choice being dependent on a variety of factors, including the purpose of the hole and the type of formation to be drilled; all these factors will influence the rate of drilling and overall cost, as will other factors such as the proficiency of the drilling team, the state of repair of the rig, the choice of rig, drilling bits, drilling fluid, etc. Rotary drilling is basically of two main types—*rotary core drilling*, wherein a core of the material is recovered to allow of precise logging of the strata and provide material for physical testing; and *rotary probe drilling*, wherein open-hole methods are employed to obtain rapid penetration. Probe drilling can be subdivided into *full-face diamond bit drilling*, *percussion drilling* and *rock bit drilling*.

rotary drilling optimisation The minimisation of the cost of a *borehole* per unit of depth. The total cost includes the capital costs of the drilling equipment and in the case of, say, a water well, the well production equipment, plus the intangible costs such as supervision, *mud* control and choice of equipment.

rotary percussive drilling A *rotary probe drilling* method borrowed from the quarrying industry and generally the cheapest. Penetration is obtained by repetitive hammering of the rock by a slowly rotating hard metal chisel, cross-chisel bit or **button bit**. In this way the full face of the hole is cut and cuttings are blown back to the surface by a high-pressure air flush. Cuttings of rock obtained are very small and difficult to log. The method is noisy, can give rise to dust problems and is least effective in sticky cohesive soils. *See also* **cable-tool drilling**.

rotary probe drilling   *Open-hole drilling* used to obtain rapid low-cost boring when detailed information on the rock is not required. A particularly useful method for probing ground where cavities exist (e.g. mines) and for drilling holes for grout injection to fill the cavities. Air flush methods are preferred, since a high up-hole velocity ensures that cuttings are recovered from the material being drilled at the time and also enables easy identification of any groundwater encountered. In some circumstances air is used to form aerated drilling fluids such as water mists, foams and aerated **muds** primarily to increase the penetration rate by reducing the differential pressure between the drilling fluid column pressure and the rock pore-water pressure. *See also full face diamond bit drilling; rock bit drilling; rotary percussive drilling*.

rotational slip   A form of instability *(Figure R.5)* usually associated with uniform cohesive soils or heavily jointed or structureless rock.

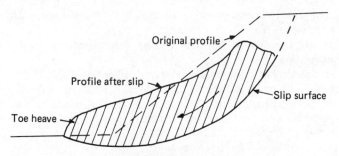

Figure R.5   Rotational slip

The slip results when the resisting forces acting along the potential slip surface are exceeded by the gravitational forces acting on the mass of material above the slip surface, causing a bodily downward rotational movement of a generally spoon-shaped mass of soil. This is accompanied by an upward and forward movement of soil at the toe of the slip. The slip surface may be circular or non-circular, depending on the uniformity of the soil conditions. Multiple rotational slips occur due to an initial local slip which triggers off a series of up-slope slips along a common basal failure plane. *'Bottleneck' slides* in the quick clays of Scandinavia and Canada are generally of this type, where local slips adjacent to unstable river

201

banks cause the clay to become liquefied, which then flows down out of the slide area and causes progressive up-slope failure as toe support is removed. A characteristic 'bottleneck' shape is produced due to the width of the slide increasing away from the river. Various methods of analysis exist for determining the safe slopes of embankments and cuttings—e.g. Taylor's method for short-term or end-of-construction case (*see Taylor's stability numbers for earth slopes*) and the *Bishop method* or *Bishop and Morgenstern method of slope stability analysis* for long-term stability.

roughneck A member of a drilling crew who works on the derrick floor.

roughness of discontinuity One of the ten parameters selected to describe discontinuities in rock masses, being the inherent surface roughness and waviness relative to the mean plane of a discontinuity. Both roughness and waviness contribute to the shear strength. Large-scale waviness may also alter the dip locally.

Reference: International Society for Rock Mechanics. Commission on Standardisation of Laboratory and Field Tests, 1977.

Rowe consolidation cell Equipment allowing *consolidation* tests to be carried out in the laboratory on samples up to 250 mm diameter, the large size allowing a specimen to be potentially more representative of the soil.

Reference: 'A new consolidation cell', P. W. Rowe and L. Barden, *Géotechnique*, **16**, No. 2, 1966.

rubber balloon method for field density determinations *See field density tests*.

rubber-tyred roller *See pneumatic-tyred roller*.

rugosity The irregularity of a *borehole* wall.

runner A timbering term, defined as a vertical member used to support the sides or face of an excavation and progressively driven or lowered as the excavation proceeds, its lower end being kept below the bottom of the excavation.

Reference: CP 2004 : 1972.

running sand Sand which is normally stable but becomes unstable in an excavation below the water table owing to *groundwater* flow carrying it into the excavation.

Ryznar stability index A general indicator for the corrosive–incrustative nature of *groundwater* and a modified form of the *Langelier index, or saturation index*, which for water within a pH range of 6.5–9.5 is the pH at which water reaches equilibrium with calcium carbonate. A stability index of less than 7.0 indicates the possibility of scale formation or incrustation, while 7.0 and above indicates increasingly severe corrosion.

Reference: 'A new index for determining the amount of $CaCO_3$ scale formed by water', J. W. Ryznar, *AWWA Jl*, **36**, 1944.

S

sabkha   A saline soil encountered in coastal salt marshes around the Arabian Gulf. Similar deposits around the world are known by a variety of names, such as: *playa*—an ephemeral lake flat; *salt playa*—as playa, but with a salty surface due to evaporation of salty lake waters; *salina*—a local depression with a high salt water table and capillary rise reaching the surface to give the formation of a salt crust.

Reference: *Arabian Salt-bearing Soil (Sabkha) as an Engineering Material*, C. I. Ellis, UK TRRL Report LR523, 1973.

sacrificial anodes   *See cathodic protection.*

safe bearing capacity   *See allowable bearing pressure; bearing capacity.*

safe yield   The rate at which water can be withdrawn from an *aquifer* without causing a long-term decline in the *water table* or *piezometric surface*. It is normally equal to the average replenishment rate of the aquifer but may be less in coastal areas where it is necessary to avoid intrusion of sea-water into the aquifer.

salina   *See sabkha.*

salinity   The proportion by weight of dissolved solids in sea-water, defined as the weight (g) of solid material dissolved in 1 kg of sea-water, the bromine and iodine having been replaced by chlorine, the carbonate converted to oxide and all organic matter oxidised.

salinometer   A marine instrument for measuring the *salinity* and temperature of sea-water.

salt playa   *See sabkha.*

saltings   Areas covered by high tidal waters, which normally have a covering of grass.

sample area ratio   *See area ratio.*

sand   The fraction of soil composed of particles between the sizes 2.0 mm and 0.06 mm. The sand may be subdivided as follows:

Grading	Particle size	British Standard sieve size to be used for separation
Coarse sand	2.0–0.6 mm	2.0 mm–600 μm
Medium sand	0.6–0.2 mm	600–212 μm
Fine sand	0.2–0.06 mm	212–63 μm

Sand is a natural sediment which is non-cohesive when dry and comprises mainly granular silicious wind or water-borne fragments derived from the products of rock weathering. *See particle size.*

203

sand blast test An *abrasion test* to measure the resistance of rocks to wear, wherein the surface of the test sample is abraded by a blast of air containing silica sand or aluminium oxide under specified conditions. The weight loss gives a measure of the rock's abrasive resistance. *See Los Angeles abrasion test.*

sand boil Sand boil is defined in relation to liquefaction as an ejection of sand and water caused by piping from a zone of excess pore pressure within a soil mass. Sand boils commonly form during or immediately after earthquakes as pressures are relieved from liquefied or other zones of excess pore pressures in subsurface saturated cohesionlesss soils. The term 'sand boil' is preferred to 'sand blow' because the latter is used to describe other phenomena such as the denudation of a local area by wind action.

Reference: 'Definition of terms related to liquefaction', *ASCE Jl Geotech. Eng. Div.*, **104**, Pt GT9, September 1978.

sand drains Vertical holes filled with sand to facilitate vertical drainage of stratified soils and relief of pore pressures in compressible soils.

sand isles Artificial structures formed by filling inside an impervious membrane with *sand*. Stability of the sand, which would otherwise slump to a very flat angle, is achieved by extracting the water from inside the membrane by pumping. The confining hydrostatic pressure is about twice the lateral pressure exerted by the sand and thus almost vertical side slopes can be maintained. The system is proposed for a variety of marine uses, including industrial and oil production islands, breakwaters and well-head protectors.

sand replacement method for field density determination *See field density tests.*

Santos constant *(a)* A granulometric term where, in a *sieve analysis*,

$$a = \frac{\Sigma y}{100n}$$

Σy being the sum of the percentages of material passing each of a set of n sieves. The set normally comprises US sieve nos. 8, 16, 30, 50, 100 and 200.

saponification The chemical process of forming a soap; more particularly, a deterioration by softening of paint films caused by the action of aqueous alkali on fatty acid constituents of the film.

Reference: CP 1021 : 1979.

saturated unit weight The weight of a material per unit volume when all the voids are filled with water.

saturation limit The water content of a soil at which further water on its surface ceases to be absorbed. The saturation level is related to the soil's *liquid limit*, being, for example, about one-half this

204

value for inorganic clays of medium *plasticity*. It is relevant to trafficability for vehicles on unsurfaced areas—e.g. *earth roads*.

saturation line, or zero air-voids line A line showing the *dry density–moisture content* relationship for soil containing no air voids. The saturation line is shown in *Figure A.3*, and is obtained by putting $V_a = 0$ in the general equation given below for the *air voids line*:

$$\rho_d = \rho_w \, \frac{1 - \dfrac{V_a}{100}}{\dfrac{1}{G_s} + \dfrac{w}{100}}$$

where ρ_d = the dry density of the soil; ρ_w = the density of water; V_a = the volume of air voids in the soil expressed as a percentage of the total volume of the soil; G_s = the *specific gravity* of the soil particles; and w = the moisture content expressed as a percentage of the mass of dry soil. Where $V_a = 0$, and ρ_w is taken as 1, then

$$\rho_d = \frac{1}{\dfrac{1}{G_s} + \dfrac{w}{100}}$$

saturated soil A soil which has its voids completely filled with water.

saturation index *See Langelier index, or saturation index.*

scanning stereoscope An instrument for binocular observation of stereoscopic pairs of photographs incorporating equipment for taking relative height measurements.

scarifier A tool comprising a number of tines set behind the front axle on a grader used to scarify the soil. The tines can be raised and lowered by the same mechanism that operates the blade.

Schlegel sheeting An impermeable flexible sheeting produced by Schlegel Engineering GmbH from high-density polythene (HDPE) and used for the containment or erosion/corrosion control of water, waste materials and industrial chemicals. Produced in rolls up to 10 m wide and 200 m long in thicknesses ranging from 1.5 mm to 3.5 mm. The sheets can be joined by fusion-welding techniques to cover large areas.

schluff A German term for soil particles of 0.06 mm size (i.e. between *silt* and fine *sand* size).

Schlumberger arrangement A linear arrangement of *electrodes* used in resistivity surveys, such that the current and potential electrodes are symmetrically placed about the centre of the spread. *See also electrical resistivity.*

Schmidt hammer An impact rebound instrument for measuring the *in situ* compressive strength of concrete and rock.

Schmidt net *See stereographic projection.*

scintillator A highly sensitive radiometric prospecting instrument for detecting gamma radiation, basically comprising a detecting element such as a thallium-activated sodium iodide crystal, optically coupled to a photomultiplier tube.

scow drilling A technique used in *cable-tool drilling* wherein a tool (scow) comprising a heavy thick-walled tube, having a hardened bevelled bottom edge and a pair of upward swinging doors a short way above the bottom, combines the cutting action of a *chisel* with the handling capacity of a *bailer (shell)*.

screw pile A pile screwed into the ground, consisting essentially of a column provided with a helix or helices at its base.

Seacalf An underwater rig operated by Fugro Ltd which can perform *cone penetration tests (CPT)* from the sea-bed in water depths up to 300 m.

SEA-FIX A high-accuracy radio position-fixing system based on the *Decca Navigator System* and similar in principle to *HI-FIX*. It provides continuous instantaneous positional control for ocean operations regardless of visibility conditions and the absence of conventional fixing facilities. Built into suitably designed lightweight buoys, the transmitting stations can be moored in the area of operation. The resultant pattern of position lines can be used relative to the buoys, and these may be fixed by celestial or other means, if so desired. *See also MINI-FIX.*

Seavane A remotely-controlled submarine vane tester marketed by Soil Instruments Ltd, London, for determining the *in situ shear strength* of soil at depth in marine boreholes.

sea-water sound velocity meter A system comprising a stainless steel underwater probe which can be lowered into the sea via a strain-bearing electrical cable connected to a surface read-out unit designed to display the speed of sound. The probe unit allows an acoustic signal of precise frequency to be reflected across a fixed distance. The time for the signal to complete its travel depends on the nature of the water, and the speed of sound is continuously calculated for each position. The equipment can be used for deep water (up to 300 m) calibration of *echo-sounder* equipment, determination of sound velocity gradients for acoustic ranging and measurement of the speed of sound through water at selected depths.

secant (interlocking) piles A system of *retaining wall* construction similar to *contiguous bored piles* but using a method of interlocking which overcomes problem of watertightness. Alternate piles are drilled and concreted and intermediate pile holes drilled, followed by chiselling a groove down the sides of the shafts. Concrete is then placed to fill both intermediate holes and grooves to form a fully interlocking and watertight wall.

secant modulus of elasticity The average slope of the stress–strain curve obtained when a material is subjected to *compression*, obtained for a range of *stress* between zero and some arbitrary value, P *(Figure S.1)*. For perfectly elastic materials, the secant

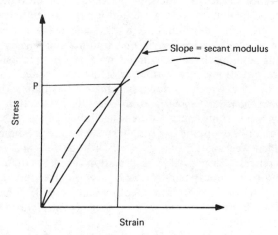

Figure S.1 Secant modulus

modulus line would lie along the stress–strain path up to the *elastic limit*. However, for some materials, such as most soils, the stress–strain path is curved along its entire path and the secant modulus is a convenient way of comparing the 'elasticity' of different soils.

secondary creep *See creep*.

sediment Material derived from pre-existing rocks by weathering and deposited elsewhere after transportation by ice, water, air or gravity.

seepage force The force due to flow, with which the seeping water acts upon the soil particles in a unit volume of soil.

seepage from discontinuity One of the ten parameters selected to describe discontinuities in rock masses, being the water flow and free moisture visible on individual discontinuities or in the rock mass as a whole.

Reference: International Society for Rock Mechanics. Commission on Standardisation of Laboratory and Field Tests, 1977.

S E gauge A gauge used for measuring *earth pressure*, consisting of two diaphragms, each having a silicon solid state *strain gauge* attached at the centre and near the edge which are connected as a Wheatstone bridge circuit.

seiches Oscillations in sea level due to the passage of an intense depression. The period between such waves may be anything from a few minutes to 2 h and the height from a few centimetres to 2–3 m.

seismic methods of surveying Seismic methods of surveying essentially comprise the introduction of seismic energy into the ground and measurement of the time taken for the generated waves to travel through the ground to a receiving station. Techniques include *continuous seismic reflection profiling* and *refraction profiling*. Although both types can be used for both on-shore and offshore situations, the former is generally limited to marine surveying, for which various systems are available, including *sparkers, boomers, pingers* and *air guns*. The refraction profiling systems use various energy sources, including gelignite explosions and the simple dropping of a weight such as a sledge hammer on an embedded steel plate.

References: 'Improved interpretation using a dual channel continuous seismic profiling system', N. C. Kelland, European Association of Exploration Geophysicists, 1970: 'The seismic refraction method—a review', R. Green, *Geoexploration*, **12**, pp. 259–284, 1974.

seismic prospecting The investigation of subsurface structure by means of *seismic waves*, using either the seismic reflection or the seismic refraction method. *See seismic methods of surveying.*

seismic velocity The velocity of *seismic waves* in a material.

seismic waves Elastic waves that are transmitted through soils and rocks as a result of disturbances such as *earthquakes* or artificial energy sources. Seismic waves are of two principal types: body waves, which travel throughout the medium; and surface waves, which are confined to the boundaries. Body waves are either compressional (*P-waves*) or longitudinal (*S-waves*). Surface waves are more complex, and include *Rayleigh waves* and *Love waves*. P-wave velocities are greater than those of other waves, and so it is the P-waves which are the first arrivals studied in refraction seismic work.

seismogram, seismograph The recorded trace from a *seismometer*.

seismology The study of the interior of the Earth by means of the global arrival times of *seismic waves* generated by *earthquakes*.

seismometer An instrument for detecting and recording vibrations of the Earth's surface such as result from the arrival of *seismic waves*.

self-potential method A method of geophysical surveying in which natural electric fields are measured at the ground surface. No external power source is applied, which distinguishes the method from that of resistivity. The electric fields arise from the electrochemical action of groundwater circulating around metallic orebodies, and the method is largely used to locate such bodies.

self-propelled vibrating rollers *See **vibrating rollers**.*

semi-diurnal tides Tides which have two high waters and two low waters in a lunar day of approximately 25 h.

sensitivity (S_t) The ratio between the undisturbed and the remoulded shear strength of a soil.

set In piling, the net distance by which the pile penetrates into the ground at each blow of the hammer.

setting All boards held in position by one frame of timber or in the case of tucking or piling frames, by two adjacent frames.
Reference: BS 6031 : 1981.

settlement Settlement is the downward movement of a structure due to compaction, elastic compression and/or consolidation of the underlying strata. The allowable settlement which a structure can tolerate without distress is dependent on both total and differential settlement, as total settlement affects the function of a building such as connections to services and appearance relative to neighbouring structures, such as tilt, and differential settlement causes stress and probable structural damage or damage to finishes. Allowable settlements of structures are reviewed by J. B. Burland and C. P. Wroth in 'Settlement of buildings and associated damage', *Proc. Conf. on settlement of structures*, Cambridge 1974, who quote the **USSR Building Code (1955)** . *Values of limiting differential settlement (see Table 14)*. Danger limits for angular distortion or relative rotation (β) of structures are given by L. Bjerrum in 'Discussion on compressibility of soils', *Proc. Europ. Conf. on Soil Mech. and Found. Eng.*, **2**, pp. 16–17, Wiesbaden, 1963. (*See* Table 13.)

The settlement of foundations on granular soils may be determined by a variety of methods, including *plate bearing tests*, *static cone penetration tests*, *standard penetration tests* and *index property tests*.

The settlement of foundations on cohesive soils is a function of the immediate elastic settlement plus the long-term *consolidation* of the material.

The *immediate settlement (P_1)* can be calculated from the expression

$$P_1 = \frac{q_n 2B(1 - m^2)}{E_d}$$

for the centre of a flexible loaded area, where q_n = the net foundation pressure; B = the width of the foundation; m = Poisson's ratio (generally taken as 0.5 for clays); and E_d = the deformation modulus. Depending on the type of foundation, depth and rigidity factors are applied.

A convenient method for estimating the average immediate settlement of a loaded area is given by Janbu, Bjerrum and Kjaernsli, N.G.I. Publication No. 16, Oslo 1956 as:

$$\delta_i = \frac{\mu_1 \mu_0 q_n B}{E_d}$$

when μ_1 and μ_0 are related to ratios of depth, length and thickness of compressible stratum to width of foundation (D/B, L/B and H/B).

The consolidation settlement (δ_c) can be calculated from

(i) $$\delta_c = \frac{\Delta e}{1 + e_0} H$$

where $\Delta e =$ the change in the *void ratio*; $e_0 =$ the initial void ratio; and $H =$ the thickness of the compressible stratum.

or (ii) $\delta_c = m_v . H . \Delta p$

where $m_v =$ the coefficient of volume compressibility; $H =$ the thickness of the compressible stratum; and $\Delta p =$ the pressure increment.

or (iii) $$\delta_c = \frac{C_c}{1 + e_0} H \log_{10} \frac{P'_0 + \Delta p}{P'_0}$$

where $C_c =$ the *compression index*; $P'_0 =$ the *effective overburden pressure*; $\Delta p =$ the pressure increment; and $H =$ the thickness of the compressible stratum.

Skempton and Bjerrum, 'A contribution to the settlement analysis of foundations on clay', *Géotechnique*, 7, 4, December 1957, proposed the use of a geological factor (μ_g) to be applied to the consolidation settlement calculation from the oedometer value to take into account the stress history of the soil.

Since consolidation settlement can take a very long time for virtual completion, it is necessary to take the useful life span of structures into account when considering allowable settlements. Thus, the degree of consolidation *(U)* that will take place in a given time must be established. This is dependent on the *time factor* *(T_v)* where

$$T_v = \frac{C_v . t}{H^2}$$

and $c_v =$ the *coefficient of consolidation*; $t =$ the time from the commencement of consolidation; and $H =$ the thickness or half-thickness of the compressible soil, depending on drainage conditions. Values of $U \sim T_v$ for various conditions of drainage are given by Janbu, Bjerrum and Kjaernsli in the paper quoted above.

Other methods of estimating settlement include the *stress path settlement analysis method*, use of centrifugal models; the Janbu deformation modulus method ('Soil compressibility as determined by oedometer and triaxial tests', N. Janbu, *Proc. Europ. Conf. on Soil Mech. and Found. Eng.*, Section 1, Wiesbaden, 1963, and 'The resistance concept applied to deformation of soils', N. Janbu, *Proc. Int. Conf. on Soil Mech. and Found. Eng.*, 1, pp. 191–196, Mexico, 1969); use of the Dutch cone sounding apparatus; and correlation with compression index for *normally consolidated* clays.

settlement gauges Instruments for measuring vertical movements of the ground and structures. Types include hydraulic, mercury and electrical, single and multiple-point gauges.

Reference: *Foundation Instrumentation*, T. H. Hanna, Trans Tech Publications, 1973.

sextant A portable marine survey reflecting instrument capable of measuring angles up to about 120° and basically comprising a telescope and a graduated arc. Two distant stations are sighted and optical images are made to coincide by rotating an arm on the graduated scale on which the included angle can be read. The position of a vessel can be fixed by taking two horizontal angles observed between three fixed shore stations.

shallow footing A footing whose width is equal to or greater than the vertical distance between ground level and the base of the footing.

shape factor *See flow net.*

shear box Equipment used both in the laboratory and in the field for the determination of *shear strength* parameters of soils. Laboratory equipment comprises either a 60 mm square box for soil containing material that will pass a 3.35 mm sieve or a 300 mm square box suitable for soil when particles pass a 37.5 mm sieve. Both types of box are split horizontally into two halves which, when moved relative to each other, cause the specimen to be sheared. Drained, undrained and consolidated tests can be carried out to give results in a similar way to those obtained using the *triaxial compression machine*. Although largely superseded by the latter, the shear box equipment has the advantage that non-cohesive material can more easily be accommodated. Also, its ability to allow of multiple reversals of the specimen makes it suitable for determining *residual and peak shear strength* parameters for use in slope stability problems. *See also Bromhead ring shear.*

shear failure A material is said to exhibit shear failure if the *shear stress* on some plane exceeds the available *shear strength* on that plane, so resulting in excessive deformation. The deformation may confine itself to the plane in question or may spread more generally throughout the material.

shear fracture   A plane on which a shear failure has occurred and has generated a *discontinuity*.

shear modulus *(G)*   The *shear stress* that gives rise to unit angular strain; hence the shear modulus is the ratio between the *shear stress* and the *shear strain. See stress–strain relationships.*

shear strain   (1) The change of the angle between two planes originally perpendicular to each other (measured in radians). (2) Shear strain is defined in relation to *liquefaction* as the change in shape, expressed by the relative change of the right angles at the corner of what was in the undeformed state an infinitesimally small rectangle or cube. As with the definition of *normal strain*, the term can be applied to deformation in cohesionless soils in either the solid or the liquefied state, and can be expressed as a percentage or in terms of a dimensionless ratio.

Reference: 'Definition of terms related to liquefaction', *ASCE Jl Geotech. Eng. Div.*, **104**, Pt GT9, September 1978.

shear strength   The shear strength of a soil is the maximum *shear stress* that the soil structure can withstand under a specified set of loading conditions and is controlled by: (a) the normal pressure on the shear plane; (b) the drainage conditions as they affect the dissipation of pore-water pressures; and (c) the rate of *strain*. The shear strength (τ_f) is a function of the *effective stress* and, in general, it has been shown that

$$\tau_f = c' + (\sigma_n - \mu) \tan \phi'$$

where $c' = $ *cohesion* and $\phi' = $ the *angle of shearing resistance* (both in terms of effective stress); $\sigma_n = $ the total normal stress; and $\mu = $ the pore-water pressure.

Shear strength may be determined by three basic categories of test:

(1) The *undrained test*, where zero dissipation of the pore-water pressure occurs during application of the normal stress and no drainage is allowed during shearing.
(2) The *consolidated undrained test*, where full consolidation is allowed under the applied normal stress but no drainage is allowed during shearing.
(3) The *drained test*, where full drainage is allowed throughout the test.

In (1) above, if pore-water pressures are measured during the test, then shear properties in terms of effective stress can be determined. If not, only the *total stress* parameters are obtainable. In (2) above, effective stress parameters can be obtained if pore-water pressures are measured. If not, then a quasi-effective stress failure-envelope is obtained. In (3) above, effective stress parameters are measured directly.

Types of tests to measure shear strength parameters include: *Bromhead ring shear; field vane test; laboratory vane test; shear box; triaxial compression machine. See also drained triaxial 'smear' test; residual and peak shear strength; soil strength.*

shear stress That component of the stress field that acts tangentially to a surface, and so gives rise to angular distortion. When written in matrix form, the shear stresses are the off diagonal elements.

shear surface, plane A continuous plane or zone on which major shear displacements have occurred. Such surfaces are often found in geological materials and are usually planes of potential weakness. Their location is thus a prime objective of a site investigation in an area believed to have been subjected to previous movement— e.g. an area of ancient landslipping.

shear wave, or S-wave *See transverse wave.*

sheepsfoot roller A machine, either towed or self-propelled, used for compaction of mainly dry fine-grained soils and provided with a hollow steel roller having rows of steel feet projecting from the surface of the rolling surface. The feet are shaped to the form of tapers, cylinders or clubs, the distribution, shape and number of which decide the performance of the roller.

sheeting Flat planks or boards (usually of timber) used to support the sides of an excavation, and kept in place themselves by a system of *walings* and *struts*.

sheet pile One of a row of piles driven or formed in the ground adjacent to one another in a continuous wall, each generally provided with a connecting joint or interlock. It is designed to resist mainly lateral forces and to reduce seepage; it may be vertical or at an inclination. *See also steel piles.*
Reference: CP 2004 : 1972.

Shelby tube sampler A thin wall open-drive sampler with integral cutting edge for obtaining undisturbed samples, generally in soft to stiff cohesive soils, and commonly used in the USA. Tube sizes are normally 2 in (50 mm) and 3 in (75 mm) diameter and 24 in (610 mm) or 30 in (760 mm) long but 4 in (100 mm), 5 in (125 mm) and 6 in (150 mm) tubes are in use.

shell and auger boring The use of shell and auger (drop-tool) or percussive rigs for boring through overburden soils is a standard method adopted for site investigation work in the UK. Basic tools include a *percussive clay cutter*, comprising a heavy open-ended steel barrel with a cutting edge or a three-vaned open cutter which is suspended on a wire line and dropped down the hole repeatedly and manually cleaned out after withdrawal from the hole; a *clay* *auger*, which causes less disturbance to the soil and is run down the hole on rods and rotated by hand; a *shell* or *bailer*, similar to a *clay cutter* but having instead a flap valve (clack) of leather or steel on the bottom of the barrel, which is 'pumped' into sand and gravel soils, raised when full and up-ended to empty while being flushed with water to clean the valve; and *chisels*, which are used to break up obstructions (e.g. boulders, etc.) or to penetrate and prove bedrock—the rock chippings being baled out with a shell. The method has the advantage of allowing accurate logging of strata to be carried out, enables *groundwater* observations and *in situ* tests such as *standard penetration tests* and *permeability* tests to be made, and allows undisturbed samples of the soils to be taken.

Shipek sediment sampler A sea-bed sampler (produced by Hydro Products, Tetra Tech Co, California, and designed by Carl J. Shipek, oceanographer) to bring virtually undisturbed unwashed samples of unconsolidated sediment from the sea-bed to the surface. It basically comprises two concentric cylinders, the inner of which is held in the open position by springs until contact is made with the sea-bed, when it rotates through 180° to engulf the sample.

shock or transient response method of pile testing A vibration test method using lightweight equipment based on a microprocessor which can measure the homogeneity of the concrete in the pile shaft and can predict pile performance. The equipment comprises a small load cell which is placed centrally on the pile shaft, and a sensitive geophone placed near the circumference of the pile head. Both are connected to a microprocessor and an oscilloscope and graph plotter, all of which are powered by a 240 V generator. The load cell is given a sharp blow with a mallet and the applied shock and velocity responses of the pile head are measured by the geophone and simultaneously measured as a fraction of time, the signals being stored in the microprocessor and displayed in the oscilloscope. From the information it is possible to deduce the pile head stiffness, pile length and cross-sectional area of the pile or, if the latter is known, then the concrete quality can be determined.
Reference: *New Civil Engineer*, 21 May 1981.

shooting or blasting Methods of well stimulation to increase *permeability* and *well efficiency* by detonating large explosive charges in the *borehole* to increase its diameter and remove fines from the borehole walls; and by causing vibratory explosions whereby a series of smaller charges are fired in sequence to produce a vibrating effect on the *casing* and formation and a surging effect due to the alternate expansion and construction of the gas bubble produced by the explosive charge.

shop drawings Drawings, diagrams, illustrations, brochures, schedules and other data prepared by a contractor, a subcontractor, a manufacturer, a supplier or a distributor which illustrate the equipment, material or some other portion of work to be executed in a contract.

Reference: *Manual of Water Well Construction Practices,* US Environmental Protection Agency, 570/9-75-001.

shore   A member (*strut*) in compression. *See dead shore; flying shore; raking shore.*

shore scleroscope   A vertical-scale scleroscope mainly used for metallurgical hardness testing but adopted for rock testing, whereby the rebound of a hammer having a diamond striking tip is related to *rock strength* and drillability.

short-bored pile   A pile normally formed by boring an uncased hole with a mobile power auger and filling with lightly reinforced concrete. It is generally used for reliable foundations in shrinkable clay which would otherwise give distress to shallow spread foundations—particularly for building on clay soils near trees.

shotcrete   *See Gunite.*

shrinkage limit (SL)   *See Atterberg limits and soil consistency.*

side boards   Timbers forming the sides of a *heading.*

side-scan sonar   A method, working on the same basic principle as an *echo sounder,* of locating objects such as rock-outcrops, pipelines, shipwrecks, etc., on the sea-bottom by emitting a fan-shaped beam of acoustic energy perpendicular to the survey craft's track, from a 'fish' towed (typically between 15 m and 150 m) above the bottom. The reflected signals are recorded as changes in density on a continuous paper roll.

side trees   Timbers supporting the *side boards* and *head trees* in a *heading.*

sidewall neutron (porosity) log   SNP (a Schlumberger trademark) is a record of epithermal neutrons, the sensing probe being mounted on a skid which is pressed against the borehole wall. *See also neutron log.*

sieve   A device for grading soils constructed either of plates with square perforations or woven wire with standard perforation size. *See also particle size; particle size distribution.*

sieve analysis   The method by which the *particle size distribution* of the *sand* and *gravel* fractions of a soil is determined. *See also mechanical analysis.*

sill, or soleplate   A member placed under the foot of a *shore* for the purpose of distributing the load.

Reference: CP 2004 : 1972.

silt   The fraction of a soil composed of particles between the sizes 0.06 mm and 0.002 mm. The silt fraction may be subdivided as follows:

215

coarse silt, 0.06–0.02 mm
medium silt, 0.02–0.006 mm
fine silt, 0.006–0.002 mm

Silt grains are gritty to the touch but barely perceptible to the naked eye. They comprise granular, mainly siliceous products of rock weathering. *See particle size.*

similar folding A pattern of folding in which successive beds in the sequence have been folded to approximately the same radius.

simple cone *See penetrometer (apparatus).*

Simplex pile A pile constructed by driving a steel tube on a detachable steel plate to the required set or penetration, the seal between shoe and *casing* being sufficient to prevent ingress of *groundwater (Figure S.2)*. After the reinforcement cage has been

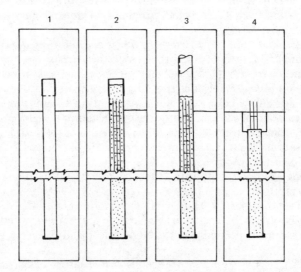

1 Tube driven to required set
2 Concrete and reinforcement (as specified) placed in tube
3 Withdrawal of driving tube and consolidation
4 Completed pile ready for capping

Figure S.2   Stages in construction of a Simplex pile

placed, the tube is filled with concrete which is continuously tamped during withdrawal. The weight of concrete plus the tamping action ensures that all voids are filled with concrete and that the latter is in direct contact with the soil. When the required *bearing capacity* cannot be obtained by a single drive, a redriven pile with an enlarged base can be used. Here the initial drive is

216

carried out to the required depth and sufficient semi-dry concrete is placed in the tube, depending on the size of enlarged base required. The rest of the tube is filled with aggregate and then withdrawn and redriven in the same position, this action forcing the aggregate into the surrounding soil and forming the concrete into an enlarged base. Subsequently, the pile is formed as for the single drive method. See also *pile foundations*.

single-acting piling hammer   See *piling hammer*.

single-packer grouting   See *double-packer grouting*.

single-roll pedestrian-controlled roller   See *vibrating rollers*.

single-tube core barrels   See *core barrel*.

Siroc grout   A grout distributed by Raymond International Inc, of New Jersey, comprising a base material, a reactant, an accelerator and water. It is economical, non-toxic and non-corrosive, and forms a permanent gel. It may be used in a single-shot batch system or a two-stream continuous system for grouting soil containing about 20 per cent or more of sand, providing high strength and low permeability.

SI system   The Système International d'Unités (SI) metric system adopted for use in the construction industry. The system includes base, supplementary and derived units.

References: *The use of the metric system in the construction industry*, PD 6031, 2nd edn, December 1978; *The use of SI units*, PD 5686 : 1972, both issued by the BSI.

skewness   If a frequency distribution deviates from a symmetric *normal distribution*, owing to the presence of a 'tail', it is said to be *skewed*. If the tail lies to the right, the distribution is positively skewed; if it lies to the left, it is negatively skewed. Skew distributions may be converted to normal distributions (normalised) by some correction to the sizing of the frequency classes. For example, the positively skewed *lognormal distribution* is normalised by adopting a logarithmic scale of size classes.

skimmer   A type of excavator fitted with a bucket that travels on a horizontal arm and can trim the ground surface. The same basic machine fitted with a jib can operate as a grab for mucking out within trenches and cofferdams, etc., or for dredging purposes.

skin friction (piling)   The contribution of the *ultimate carrying capacity* of a pile due to the force developed between the pile shaft and the surrounding ground. Where cohesive soil surrounds the pile shaft, the term *adhesion* is also used. See also *negative skin friction (piling)*.

slab and block slide   A form of *translational slide* where the mass of sliding material remains more or less intact. This can occur where a surface layer of rock overlies a plane of weakness (e.g. a clay-filled joint running roughly parallel to the ground surface) or similarly where a hard clay crust overlies softer clay.

slake durability test   A *rock mechanics* test to determine the climatic weatherability of rocks, wherein representative rock samples are placed in a perforated drum which after drying and weighing is half submerged in water and rotated at 20 rev/min for 10 min. The drum is then again dried and weighed and the percentage ratio between the dry weights before and after slaking gives the slake durability index.

sleech   A term for alluvial *silt*, used in the north of England.

sleeper walls   Dwarf walls located at intervals between the main load-bearing walls providing intermediate supports to a suspended ground floor.

slickensides   Grooves or scratches which are cut on one or both of two surfaces that have been sheared against each other. The grooves lie parallel to the direction of relative movement, and are formed by asperities in one surface ploughing into the other. Slickensides are characteristic of fine-grained soils or rocks; for example, they may be found on the boundary shear surfaces of landslips in clays, or on fault surfaces in shales or phyllites. Slickensides are often associated with surface polish due to the reorientation of platy minerals along the shear surface.

slip   *See rotational slip.*

slip indicator   A device for determining the location of a zone within a soil mass where movement is occurring. It comprises a flexible PVC tube surrounded by *sand* and fixed to a baseplate at the base of a *borehole* extending to ground level. The movement zone is determined by lowering a probe down the tube until resistance due to deformation of the tube is felt.

slope angle   The angle at which a slope is inclined to the horizontal. Expressed in degrees to the horizontal or as a percentage grade— i.e. a vertical rise of 10 m in 100 m horizontal distance can be expressed as a slope in 1 in 10, or 5.71°, or 10 per cent.

slug test   A test used in *groundwater* investigation for measurement of *aquifer* parameters, providing transmissivity, well-response time and an estimate of storativity.

Reference: 'The use of the slug test in groundwater investigations', J. H. Black, *Water Services* (*G.B.*), March 1978.

slurry trench   A trench backfilled with slurried soil to provide an impermeable cut-off wall.

smooth-wheeled roller   A machine used for compaction of fill, road formations and asphalt carpets, etc., comprising two or three large metal rollers. It is available with choice of standard transmission with clutches for general work or with torque converter or hydrostatic transmission to give smooth shockless drive. Weight distribution can be varied by addition of sand or water ballast to the rollers or by attaching weights to the frame.

soakaways Excavations, either open or infilled with coarse granular material, taken down into permeable soil, into which surface water from roads, footpaths, roofs and yards can be discharged. Design rules are given in 'Soakaways', Building Research Establishment *Digest*, 151, 1973.

soda ash, or washing soda, or sodium carbonate An alkali used to reduce the hardness of water by precipitating calcium and magnesium salts, and to reduce its acidity.

soft rock An imprecise term used to denote materials that are more coherent than soils, yet are less coherent than many rocks. Two empirical definitions are that a soft rock does not need to be excavated by blasting, and that a soft rock has an unconfined compressive strength of between 400 and $1000 \, kN/m^2$. It may be noted that to the geologist all sedimentary rocks are termed soft rocks, which may lead to confusion in an engineering context, since many sedimentary rocks cannot be excavated mechanically.

software Methods, techniques and programmes, etc., used for the interpretation of data in computers (*hardware*).

soil In the engineering sense, any of the drift deposits forming part of the Earth's crust, except for the agricultural *topsoil*, which are not part of the solid 'rock' formation. It may comprise *clays, silt, sands, gravel, cobbles* and *boulders* formed either by sedimentation of the various constituents carried into place by wind-, water- or ice-borne agencies, or formed from the parent rock *in situ* by various weathering agencies—e.g. freezing and thawing cycles, water softening or leaching out of cementitious material. *See also soil strength.*

soil assessment cone penetrometer A penetrometer developed to measure the ability of soil to carry military vehicles and comprising a 30° included angle cone 12.5 mm diameter at its base, on the end of a 600 mm long 9.5 mm diameter steel shaft. Penetration into a soil allows its *California Bearing Ratio* (range 0–15%) to be read off by a pointer operated by a calibrated compression spring.

Reference: 'The strength of clay subgrades: its measurement by a penetrometer', W. P. M. Black, UK TRRL Report 901, 1979.

soil–cement pavement A method whereby soil is mixed *in situ* or premixed with cement and compacted by rolling to form a *sub-base* for a flexible pavement. Where suitable (usually granular) soils exist, the method is economical for reasonably large pavement areas, and satisfactory sub-bases for flexible road and airfield pavements may be formed.

soil compaction The state of compaction of a soil is dependent upon its type, moisture content and amount of compacting effort applied. For a given compacting effort and soil type, a relationship (*see Figure A.3*) exists between the *dry density* (γ_d) of the soil and the *moisture content*, and a *maximum dry density* is reached at a particular moisture content known as the *optimum moisture content* (OMC). The effect of increasing the compacting effort is to increase the γ_d but to decrease the OMC. Standard laboratory tests for determining dry density–moisture content relationships are the *standard Proctor compaction test*; the *modified Proctor compaction test* and the *modified AASHO compaction test*, which are similar but use a greater compacting effort; and the vibrating hammer test, used for granular soils. The results of such tests can be used to forecast the field performance of compacting plant.

soil constitution method for determination of thermal conductivity A method wherein the thermal conductivity of the ground is assessed from laboratory tests carried out on an undisturbed sample of the soil. *See thermal conductivity of soil; transient needle method.*

soil description and soil classification There is a difference between the engineering description and the classification of soils.

A full description (*see* BS 5930:1981, Code of Practice for Site Investigations) gives detailed information on the mass characteristics, including strength, bedding, discontinuities and state of weathering, plus material characteristics, including colour, particle shape, composition, grading, plasticity and soil name. This information is normally given, for example, in the borehole records of a site investigation report, and typical descriptions would be:

stiff closely fissured grey *clay* of high plasticity (London Clay)
dense yellow fine to medium *sand* with thin lenses of soft grey silty clay (recent alluvium)

Soil classification, on the other hand, provides a concise and systematic method of grouping different types of soil by reference to their main characteristics and is primarily for soils to be used as constructional materials.

The Casagrande System of soil classification was devised in 1942 and adopted by the US Corps of Engineers for use in airfield construction. The system allowed soils to be designated by group symbols consisting of prefix and suffix letters—the prefixes indicating whether the soil was coarse (gravel/sand), fine (silty/clay) or organic and the suffixes providing quaifying terms depending on grading and plasticity—e.g.

GW well-graded gravel–sand mixture, little or no fines
GP poorly graded gravel–sand mixture, little or no fines
CL clayey silt (inorganic)

The system was subsequently extended and adopted by the UK Transport and Road Research Laboratory. In 1952 the US Bureau of Reclamation and the Corps of Engineers consulted with A. Casagrande of Harvard University and agreed further modifications which led to the Unified Soil Classification System. This took engineering properties of the soil into account. Further development led to the current British Soil Classification System (BSCS) given in BS 5930:1981, Code of Practice for Site Investigations.

The Atterberg limit tests are used in the 'plasticity chart' (*Figure S.3*) to determine the group classifications of fine-grained soils. The

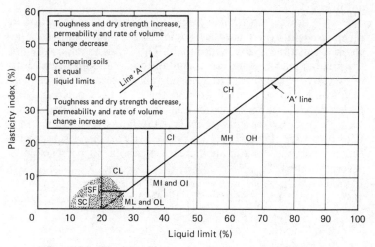

Figure S.3   Plasticity chart used in Casagrande soil classification

chart incorporates the 'A-line', which provides an empirical boundary between inorganic clays and silty and organic soils. Although the latter two types of soil overlap on the chart, they can normally be easily differentiated by visual examination.

See also AASHO soil classification system; airfield classification systems; Atterberg limits and soil consistency; unified soil classification system.

soil fabric   Soil fabric is defined by Rowe as a term referring to the size, shape and arrangement of the solid particles, the organic inclusions and the associated voids. The term 'structure' is applied to the element of fabric which deals with the arrangement of a particular size range. Thus, clay particle arrangements constitute 'structure', whereas the arrangement of particle groups (for example, in layers having different particle size) comes under 'fabric'. Reference: 'The relevance of soil fabric to site investigation practice', P. W. Rowe, 12th Rankine Lecture, *Géotechnique*, **22**, No. 2, June 1972.

soil mechanics   The investigation, description, classification, testing and analysis of soils and rocks (generally limited to weathered rock) to determine their inter-reaction with structures built on, against or with them. *See rock mechanics.*

soil name   A name based on *particle size distribution* and plastic properties, such as *gravel, sand, silt* or *clay.*

soil section   A diagram showing the soil stratification along a given line and usually relative to a particular datum—e.g. ordnance datum.

soil set-up   'Soil set-up' or '*pile freeze*' is the temporary reduction in *shear strength* experienced by some semi-cohesive soils such as chalk when a pile is driven into them. This is due to remoulding of the soil, the release of moisture from the soil grains and a build-up of excess pore pressure. The soil resistance increases with time as reconsolidation of the soil around the pile takes place and the excess pore pressures dissipate. Vijayvergiya *et al.*, in 'Effect of soil set-up on pile driveability in chalk' (*ASCE Jl Geotech. Eng. Div.*, GT 10, October 1977) report a rapid increase of soil resistance with time, reaching a maximum increase of about 80 per cent after several months.

soil strength   The following scales of soil strength are given in the British Code of Practice for Site Investigations (BS 5930: 1981).

(a) *Cohesive soil*

Term	Shear strength	Field test
Very stiff	>150	Brittle or very tough
Stiff	75–150	Cannot be moulded by fingers
Firm	40–75	Can be moulded by strong pressure in fingers
Soft	20–40	Easily moulded in fingers
Very soft	<20	Exudes between fingers when squeezed in fist

(b) *Non-cohesive soil*

Relative density		Field test	
Term	Standard penetration 'N'-value	Term	Test
Very loose	<4	Loose	Can be excavated with a spade: 50 mm wooden peg can be driven easily
Loose	4–10	Dense	Required pick for excavation:
Medium-dense	10–30		50 mm wooden peg hard to drive
Dense	30–50	Slightly cemented }	Pick removes soil in lumps which can be abraded
Very dense	>50		

In a modified scheme for Piteau (*see Table 2*) an intermediate term is given covering the overlapping range for hard cohesive soil and very weak rock, the shear strength for which is given as 0.300–0.625 MN/m^2 (unconfined compressive strength 0.60–1.25 MN/m^2). *See also* **rock strength**.

Reference: 'Geological factors significant to the stability of slopes cut in rock', D. R. Piteau, *Proc. Symp. Planning Open Pit Mines*, Johannesburg, 1970.

soil structure *See soil fabric.*

soil suction The pressure difference or reduced pressure of 'held' water in soil above the *water table* and a measure of the ability of the soil to suck in moisture until the pressure in the soil pores equilibrates with that outside. Suction is directly related to pore size and, hence *capillarity* and may be measured by tensiometer, suction or pressure plate, pressure membrane apparatus or vapour absorption methods.

soldiers Vertical members supporting horizontal *poling boards* or *walings*.

Reference: CP 2004 : 1972.

soleplate *See sill, or soleplate.*

solid core recovery (SCR) A parameter used in the description of rock core recovered from a *borehole*, and defined as the length of core recovered as solid cylinders, expressed as a percentage of the length drilled.

solifluction A style of rapid mass movement associated with the production of meltwater during the thaw period in cold environments. The meltwater is normally produced by the thawing of ground ice, but some contribution from snow melt can also occur. Movement takes place on a shallow shear plane whose depth is limited by the depth of the *active layer*, and is assisted by the production of excess pore pressures as a result of the process of *thaw consolidation*. Solifluction affects mainly fine-grained soils, and shallow, relict shear surfaces in clay soils in Britain have been attributed to solifluction at the close of the last Ice Age. *See also gelifluction*.

solifluction lobe The morphological feature that results from soil movement by *solifluction*. The lobes are usually small features a few metres across, and have margins accentuated by banks of coarse stones or ploughed-up vegetation.

solifluction sheet A layer of soil that has moved into position by *solifluction*. In detail, solifluction sheets are the product of many contiguous *solifluction lobes*. The material of which the sheet is composed is generally closely related to local bedrock sources, but is poorly compacted and contains angular fragments set in a fine matrix. It is sometimes difficult to tell soliflucted debris from *till*, but possible criteria include the lack of far-travelled (extrabasinal) material and the lower density of the soliflucted material. Solifluction sheets may be traced for considerable distances, and are capable of moving on very shallow slopes, owing to the generation of excess pore pressures due to *thaw consolidation*.

sonic log Also called *acoustic-velocity log* or *continuous velocity log*, the sonic log gives the time for acoustic waves to travel a set distance. It is the reciprocal of longitudinal (*P-wave*) velocity and is used for *porosity* determinations (*see Wyllie relationship*). To compensate for errors due to tilting of the sonde and changes in hole diameter, the results are averaged from two alternately pulsing sonic transmitters.

sound Vibrations of the air perceptible to the ear and giving rise to the sensation of hearing. *See also decibel; equivalent continuous sound level; noise; reflected sound; sound attenuation; sound level; total sound level*.

sound attenuation The reduction of sound with time. Where a noise has been measured as giving a certain *equivalent continuous sound level (L_{eq})* over a period T, then each halving of that period gives a reduction of 3 dB(A) to the L_{eq} for that noise source, i.e. an L_{eq12} of 90 dB(A) over 12 h would be 87 dB(A) if operated for 6 h, and 84 dB(A) if operated for 3 h, etc., with no allowance for reflection, absorption, wind, etc. A noise level measured at a given distance from the source reduces by 6 dB(A) with each doubling of distance, i.e. a level of 90 dB(A) at 10 m becomes 84 dB(A) at 20 m, 78 dB(A) at 40 m, 72 dB(A) at 80 m, etc. N.B. A theoretical maximum source noise is the *sound level* at 10 m plus 28 dB(A) or the sound level at 1 m plus 8 dB(A). *See also decibel; reflected sound; sound; total sound level*.

sound level The amount of sound (or noise). This is normally measured by an electrical meter containing built-in frequency weightings. The usual weighting met in Civil Engineering is the 'A', weighting which is designed to make the instrument respond more closely to the response of the human ear. Sound is measured in decibels (dB) and 'A' weighted sound is recorded as dB(A). Sound level in dB:

$$L_p = 20 \log_{10}\left(\frac{p}{p_0}\right)$$

where p = r.m.s. sound pressure in N/m^2; and p_0 = reference level in N/m^2, normally taken in air at the threshhold of hearing or at $2 \times 10^{-5} N/m^2$.
NB. A theoretical maximum source noise is the sound level at 10 m $+28$ dB(A) or sound level at 1 m $+8$ dB(A).
See also **equivalent continuous sound level; reflected sound; sound; sound attenuation.**

sounding A method of determining water depth by lowering a calibrated chain or lead line on to the sea- or river-bed. Soundings should preferably be taken at slack water with a minimum of three readings, and the time recorded so that reference can be made to the water level recorded at these times on the nearest tide gauge. In this way the bed levels may be related to datum level.

spacing of discontinuity One of the ten parameters selected to describe discontinuities in rock masses, being the perpendicular distance between adjacent discontinuities; it normally refers to the mean or modal spacing of a set of joint.
Reference: International Society for Rock Mechanics. Commission on Standardisation of Laboratory and Field Tests, 1977.

sparkers Instruments used in marine geophysical surveying which produce an electrical discharge as a seismic energy source. *See* **continuous seismic reflection profiling.**

special Franki *See* **Franki piling systems.**

specific absorption A measure of the capacity of water-bearing material to absorb water after all gravity water has been removed, and defined as the ratio between the volume of absorbed water and the volume of saturated material.

specific capacity of a well The flow (discharge) of water divided by the **drawdown** from the original piezometric surface or groundwater level. Water level in a **well** tends to become lower with increased pumping, and specific capacity will tend to decrease with time.

specific conductance The electrical conductivity of water when at a temperature of 25°C, measured in micro-ohms per centimetre.

specific driving resistance A value *(r)* related to the energy of a hammer required to drive penetration test equipment into the ground:

$$r = N \frac{Wh}{eA}$$

where N = the number of blows; W = the weight of the hammer; A = the cross-sectional area of the test equipment; h = the height of the drop; and e = the depth of penetration. For the **standard penetration test**, $r \simeq 770N$ kN/m^2.

specific energy   The work done in removing a unit volume of rock by **rotary drilling**, given by

$$e = \frac{F}{A} + \frac{2\pi NT}{AR}$$

where F = thrust; N = rotary speed; R = penetration rate; and T = torque. The term F/A is small compared with the torque component and e reduces to $2\pi NT/(AR)$.

Reference: 'CIRIA instrumented drilling trials—background and progress', E. T. Brown, *Ground Engineering*, January 1979.

specific gravity   The ratio between the mass of a material and the mass of an equal volume of water at a temperature of 4° C. The specific gravity (s.g.) of most soils is approximately 2.65. However, it is sometimes important to determine the value more accurately— e.g. in the calculation of **void ratio** of a soil, in the determination of **moisture content** by the **pycnometer** method, and for **particle size** analyses when methods such as the 'density bottle', 'pycnometer' or immersion using Archimedes' principle (for large rock fragments) are used.

specific heat   The ratio between the amount of heat required to raise the temperature of a given weight of material through 1 degree Celsius and that required to raise an equal weight of water through 1 degree Celsius.

specific resistance   The resistance between opposite faces of a unit cube of a given material at a given temperature.

specific retention   The ratio between the amount of water that a given volume of soil or rock will retain against gravitational pull and the total volume of that material.

specific storage   The volume of water released from or taken into storage per unit volume of an aquifer per unit change of head.

specific surface   The ratio, for a given weight of a substance, between the surface area and its volume.

specific yield   The ratio between the amount of water that a given volume of soil or rock will yield under gravitational pull, to the total volume of that material.

specification The instructions given on the contract documents to bidders, consisting of the general and special conditions for execution of the contract together with the technical provisions.

spectral log A well log that records both the energy and the intensity of *gamma rays*.

spectrophotometer An optical instrument with which the colour intensities of spectra may be compared.

speedy moisture content tester A moisture meter in which calcium carbide is used to generate acetylene in a closed container connected to a pressure gauge. The pressure developed is proportional to the moisture present in the soil sample.

Spencer method of slope stability analysis This method of analysis for the stability of earth embankments assumes a cylindrical failure surface and incorporates the method of slices based on work by Fellenius (1927) and Bishop (1955), and the *pore-pressure ratio (r_u)* considered by Bishop and Morgenstern in 1960. Charts are given for a range of stability factors, N_s, against slope angle, β, and ϕ'_m (the mobilised friction angle whose tangent is $\tan \phi'/F$), for various values of r_u. N_s is defined as

$$\frac{c'}{F\gamma H}$$

where c' = the effective cohesion; F = the factor of safety = available shear strength/mobilised shear strength; γ = the soil density; and H = the height of the embankment.

Reference: 'A method of analysis of the stability of embankments assuming parallel inter-slice forces', E. Spencer, *Géotechnique*, **XVII**, No. 1, pp. 11–26, 1967.

spherical coordinates A three-dimensional coordinate system based on a series of concentric spheres, each sphere carrying lines of latitude and longitude. Thus, a set of spherical coordinates will be of the form (r, θ, ϕ) where r is the radial distance to the origin, θ is the azimuth angle in the equatorial plane and ϕ is the elevation relative to the equator.

spherical stress *See hydrostatic stress.*

sphericity The ratio between the surface area of a sphere having the same volume as a particle and the surface area of that particle.

SP log (spontaneous potential, or self-potential) The difference in potential measured between an *electrode* which is lowered down a *borehole* and one which remains fixed at the top of the hole. Measurements are normally taken to provide a continuous log throughout the depth of the hole. The SP log is a common method of down-hole logging in the oil industry and increasingly so in the investigation of fresh-water aquifers. Extensive logging of oil wells has established relationships between various parameters such as SP, resistivity and ionic concentrations for oil-field brines which have shown

$$\mathrm{SP} = K \log \frac{R_{\mathrm{mf}}}{R_{\mathrm{w}}}$$

where K = the formation factor; R_{mf} = the resistivity of the *mud* filtrate; and R_{w} = the resistivity of formation water. This relationship can be misleading when applied to freshwater sands for *aquifer* determination, for which reference may be made to Alger. Reference: 'Interpretation of electric logs in fresh-water wells in unconsolidated formations', R. P. Alger, *Soc. Professional Well Log Analysts' Symposium*, Houston, 1966.

spoil Excavated soil, rock or other material which is surplus to requirement and is generally removed from the site.

spontaneous liquefaction A phenomenon that takes place in loose saturated silty sands due to high pore pressures produced by vibrations from seismic, blasting and even pile-driving activities. It can occur on virtually flat ground and has been known to spread hundreds of kilometers at speeds of over 100 km/h—e.g. at Grand Banks, Newfoundland, in 1929. *See also* **liquefaction**.

spring-type gravimeter *See gravimeter*.

springs Natural discharges of groundwater at ground level due to the down-gradient parts of an *aquifer* being exposed at the surface. Meinzer classified the discharge of springs as follows:

Order	Discharge (m^3/day)
1st	> 245 000
2nd	24 500–245 000
3rd	2450–24 500
4th	545–2450
5th	54–545
6th	5.4–54
7th	1.3–5.4
8th	< 1.3

Reference: 'Outline of groundwater hydrology with definitions', E. O. Meinzer, US Geol. Survey Water Supply Paper 494, 1923.

spring tides The two occasions in a cycle of one lunar month (approximately $29\frac{1}{2}$ days) when the average range of two successive tides is greatest.

squeeze cementing The action of forcing cement slurry into place by high-pressure pumping.

228

SRO Special road oil, comprising creosote cut-back bitumen, used in the *wet sand process* of stabilisation.

stage method of grouting *See curtain grouting.*

standard drilling *See cable-tool drilling.*

standard hydrogen electrode A reference *electrode* comprising an electro-positive metal, such as platinum, in an *electrolyte* containing hydrogen ions at unit activity and saturated with hydrogen gas at 1 standard atmosphere.
Reference: CP 1021 : 1979.

standard penetration test This test consists of driving a 50 mm (2 in) diameter split spoon (barrel) sampler weighing 66.7 N (15 lbf) into the soil at the bottom of borehole by means of an *automatic trip hammer* which allows a 622.7 N (63.5 kgf) (140 lbf) weight to drop freely through a height of 760 mm (30 in). The number of blows to give a penetration of 300 mm (12 in) after the tool has been driven through an initial 'seating' depth of 150 mm (6 in) to penetrate a possible disturbed zone caused by the boring operations is termed the penetration resistance (N). In gravel and weak rocks the driving shoe is replaced by a 50 mm (2 in) diameter, 60 degree solid cone and the test is then known as the *standard penetration test (cone)*. The test was initially introduced by The Raymond Concrete Pile Co, USA, to provide data for the prediction of the carrying capacity of piles in sand. It is now a standard field test in site investigations for use mainly in granular soils, but also for clays and weak rocks, and many researchers have enabled the results obtained to be extended to allow various soil parameters used in foundation design to be predicted with reasonable accuracy.

Terzaghi and Peck gave the following relationship for relative density of sands based on the standard penetration test value (N):

No. of blows	Density of sand
0–4	Very loose
4–10	Loose
10–30	Medium-dense
30–50	Dense
> 50	Very dense

In addition, a chart was given relating N-values to allowable bearing pressures for various width footings placed in sand. Subsequent experience indicated that this gave conservative values, and current practice involves modification of the field value of N using a graphical relationship by Thorburn which takes into account the effect of overburden pressures, principally researched

229

by Gibbs and Holtz. Stroud concluded that simple relationships existed for insensitive clays and soft rocks between N and mass shear strength (C_u) of the form $C_u = f_1 N$ and between N and the coefficient of volume compressibility (m_v) of the form

$$m_v = \frac{1}{f_2 N}$$

where f_1 and f_2 are constants.

References: *Soil Mechanics in Engineering Practice*, K. Terzaghi and R. B. Peck, Wiley, New York, 1948; 'Tentative correction chart for the penetration test in non-cohesive soils', S. Thorburn, *Civ. Eng. and Public Wks. Rev.*, **58**, No. 683, 1963; 'Penetration testing in the United Kingdom', S. Rodin, B. O. Corbett, D. E. Sherwood and S. Thorburn, *Proc. Europ. Sym. on Penetration Testing*, Stockholm, June 1974, **1**, pp. 139–146; 'Research in determining density of sands by spoon penetration testing', H. J. Gibbs and W. G. Holtz, *Proc. 4th Int. Conf. on Soil Mech. and Found. Eng.*, **1**, London, 1957; 'The standard penetration test in insensitive clays and soft rocks', M. A. Stroud, *Proc. Europ. Sym. on Penetration Testing*, Stockholm, June, 1974.

standard penetration test (cone)   *See standard penetration test.*

standard Proctor compaction test   A test whereby soil is compacted into a special mould using a 5.5 lb rammer with a 12 in drop at various moisture contents. The resulting densities are plotted against moisture content *(see Figure A.3)*, from which plot the *optimum moisture content* and *maximum dry density* values for the soil are obtained. From these results the field compaction of the material as fill can be specified. The test is also known as AASHO Test T-99, ASTM method D698-70.A and BS Compaction Test ($5\frac{1}{2}$ lb (2.5 kg) rammer method—BS 1377 : 1975, Test 12).

standard U(100) soil sampler   A sampler used in the UK for routine use in site investigations with cable percussion/shell and auger boring, to obtain undisturbed samples of cohesive soils and soft rocks such as chalk and weathered Keuper Marl. It comprises a steel barrel about 450 mm long by 100 mm internal diameter to which a cutting shoe and a drive head can be screwed. After sampling, the tube can be sealed after waxing the sample, with screw or push-on end caps. Has an area ratio of about 30 per cent. Formerly, commonly called U(4) sampler.

standpipe (piezometer)   *See piezometer.*

state of the art   The most up-to-date knowledge of a particular subject currently being put to practical use.

static–dynamic probing   A method of probing using the *Dutch sounding test* by static loading where soil conditions permit but using an *automatic trip hammer or monkey* as used in the *standard* *penetration test* to obtain penetration of the cone through stronger materials.

static cone penetration test *See cone penetration test (CPT).*

static penetration test *See cone penetration test (CPT).*

static point resistance (q_c) The average pressure acting on the conical point in the standard static *cone penetration test (CPT).* *See also Dutch sounding test.*

static probing test *See Dutch sounding test.*

station pointer An instrument used for plotting a resection fix from a pair of sextant angles taken on three shore stations (i.e. with a common centre mark). It consists of three radiating arms which can slide over a circular graduated scale *(Figure S.4)*. The arms are set

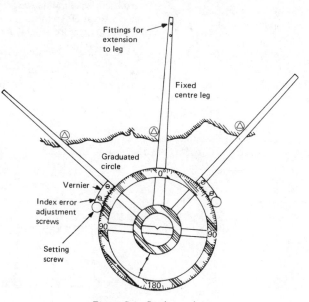

Figure S.4 Station pointer

to observe angles and fitted so as to pass through the plotted positions of the stations, thus giving the location of the observer at the centre notch of the station pointer.

steel piles Steel interlocked sections *(see Figure S.5)* are widely used to form *sheet pile* walls for retaining structures. Special *box piles* can be incorporated in the wall to enable it to carry vertical loading and universal beams similarly incorporated to allow the wall to resist high bending stresses. Steel bearing piles, including box, tubular and H-section universal bearing piles *(see Figure S.6)* are

231

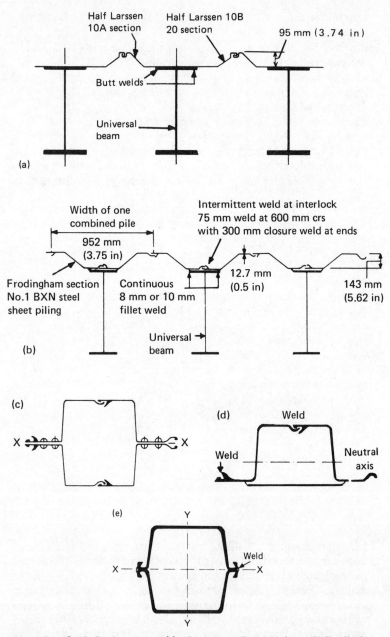

Figure S.5 Steel piles incorporated in sheet pile walls: (a) Unissen, (b) Frodingham high modulus, (c) Frodingham double box, (d) Frodingham plated box, and (e) Larssen box piles

232

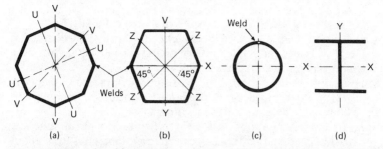

Figure S.6 Examples of free-standing steel piles: (a) Frodingham box, (b) Rendhex box, (c) tubular, and (d) H-section

used in the construction of piers, jetties, bridges and buildings. Steel piles can withstand heavy driving into hard ground, offer high resistance to horizontal forces (particularly useful in resisting impact berthing forces in marine construction) and can easily be extended or shortened on site by welding or torch-cutting as necessary.

stemming The plugging of joints or openings in timbering or sheet piling with hay, straw or similar material to prevent the loss of retained soil such as sand.

step taper piles *See Raymond piling systems.*

stereo pair Two photographs of the same subject taken from adjacent viewpoints such that when viewed simultaneously through a suitable viewer, the images in the two photos can be fused into one apparently three-dimensional image. The term is mostly applied to pairs of aerial photographs with an area of overlap for viewing in the above manner.

stereographic projection A construction that enables three-dimensional data on the orientation of planes or lines to be displayed on a diagram. The plane or line is considered to lie at the centre of a hemisphere, and its projection into the surface of that hemisphere is displayed. Planes and lines will project as great circles and points (poles), respectively. The orientation is specified in terms of lines of latitude and longitude on the surface of the hemisphere, which are also projected. Two base diagrams are in common use: the *Wulff net*, which is a direct projection of the lines of latitude and longitude, and the equal-area or *Schmidt net*, in which all equal areas on the hemisphere are projected as equal areas on the net. The former type of net preserves geometric relationships and allows three-dimensional problems in solid geometry to be solved graphically. This is used in the analysis of geological structures. The latter type of net preserves the density of projected points and is used in problems where the strength or clustering of orientations are relevant. Points on such a projection are often contoured and subjected to statistical analysis.

sticky limit The lowest water content at which soil will stick to metal. It is a useful guide in determining the allowable operating conditions for metal-tracked vehicles.

Stingray A remote-controlled, hydraulically-operated, seafloor-based unit developed by McClelland Engineers for carrying out offshore *in situ* testing and soil sampling.

Stokes's law A relationship for the settling velocity of spherical particles in liquid. It is applicable for particle diameters from about 0.2 to 0.0002 mm for which lamina flow past the particles applies, as larger particles cause turbulent flow and *Brownian movement* occurs with smaller particles. It is used as a basis for sedimentation analysis for soil particles smaller than about 0.07 mm equivalent diameter (i.e. silty/clay size). The fact that soil grains are not truly spherical is not of practical importance regarding the accuracy of the results. Stokes's law states

$$\text{terminal velocity } (v) = 2.18 \, g \,.\, 10^{-2} \,.\, \frac{r^2}{\eta} \, (\gamma_s - \gamma_w)$$

where g is the gravitational constant; v is in m/s; r = the particle radius (mm); η = the viscosity of the liquid (N s/m^2); γ_s is the density of the particle (Mg/m^3); and γ_w is the density of water (Mg/m^3).

stoop and room *See subsidence.*

storage coefficient The volume of water released from storage or taken into storage, per unit surface area of *aquifer*, per unit change in head. The term is synonymous with *specific yield* for *unconfined aquifers*.

strain Change of shape as a result of an applied stress.

strain gauge A device designed to record microstrain movements by measuring small changes in electrical output when fixed or bonded in some way to the equipment under load. Resistance strain gauges make use of the fact that a change in resistance occurs as the gauge deforms under load. The earlier type of such gauges used carbon rods or films of carbon on flexible backings, and more recent types use fine resistance wire bonded onto paper or foil types with printed circuits. Other types include semiconductor gauges, comprising a single element cut from a crystal of semiconducting material (e.g. silicon or germanium), and photoelastic and piezoelectric gauges.

stratum contour *See strike line.*

stress The force that produces deformation of a body. It is measured in units of force per unit area.

stress-path settlement analysis A method of settlement analysis whereby the *effective stress* changes likely to occur during loading of the soil in the field are first computed and the resultant anticipated stress path is applied during *oedometer* tests on soil samples in order that the vertical strains approximate as closely as possible to the likely in situ conditions. Ideally, the stresses are computed for different depths below foundation level and oedometer tests made on samples from these depths are subjected to the relevant stress-path patterns. The vertical strains are then summed to give the predicted *consolidation* settlement. In the same way, the immediate undrained settlement (δ_1) is computed by taking

$$\delta_1 = \frac{\Delta\sigma_v - \Delta\sigma_h}{E(z)} \, dz$$

where $\Delta\sigma$, and $\Delta\sigma_h$ are increments in total vertical and horizontal stresses at depth z taking into account the changes in vertical and horizontal stresses, and E is Young's modulus.

stress–strain relationships In a linear homogeneous isotropic material the following relationships exist between the elastic parameters:

shear modulus $(G) = \dfrac{E}{2(1+v)}$

Lamé constant $(\lambda) = \dfrac{Ev}{(1+v)(1-2v)}$ where $\dfrac{\lambda}{G} = \dfrac{2v}{1-2v}$

bulk modulus $(K) = \dfrac{E}{3(1-2v)} = \dfrac{2(1+v)G}{3(1-2v)}$

Young's modulus $(E) = \dfrac{9KG}{3K+G}$

Poisson's ratio $(v) = \dfrac{(3K-2G)}{2(3K+G)}$

constrained modulus (in soil mechanics) $\left(\dfrac{1}{mv}\right) = \dfrac{(1-v)E}{(1+v)(1-2v)}$

Reference: *Elastic Solutions for Soil and Rock Mechanics*, H. G. Poulos and E. H. Davis, Wiley, 1973.

strike direction The direction along which a geological stratum retains a constant elevation. This direction is perpendicular to the direction of true dip.

strike line A contour line drawn on a geological horizon such as the contact between two beds, a fault plane, etc. If the surface is planar, the strike lines are a series of uniformly spaced parallel lines that run in the *strike direction* of the horizon.

strip foundation A foundation providing a continuous longitudinal bearing. *Narrow strip foundations* (approximately 400 mm wide) or *trench-fill foundations* are those cut by mechanical excavators and filled with mass concrete, and are used where deep strip footings of 0.9 m depth or greater are required and where the soil will stand unsupported. They result in saving in volume of excavation and brickwork over normal strips and, additionally, do not require hand trimming of sides and bottom of the trench. *Traditional strip foundations* are of a width generally governed by minimum working space required by a bricklayer and are practically about 450 mm wide when 2-skin cavity wall brickwork is placed on a 150 mm concrete bed. *Wide strip foundations* are those where width necessitates transverse reinforcement.

strut A member in compression—e.g. supporting the *walings* of a *cofferdam*.
Reference CP 2004 : 1972.

sub-base The layer in a pavement immediately below the road base. It can consist of crushed rock, slag or concrete, well-burnt non-plastic shale; natural sand and gravel; lean concrete; cement-stabilised materials; or soil cement. *See also* **pavement design**.

subcontractor An individual, firm, company or corporation having a direct contract with the contractor or with any other subcontractor for the execution of a part of the *work* at a site.

subdrains Open-jointed or perforated pipes laid in a trench in the bottom of excavations to drain the ground as the work proceeds.

subgelisol The unfrozen zone below the *permafrost*.

subglacial tills *See till*.

subgrade The soil formation on which the road pavement is constructed. *See also* **pavement design**.

submerged unit weight The total unit weight less the unit weight of water.

submersible pump A pumping unit that has the element and the motor submerged in the liquid to be pumped.

subsidence The downward, mainly vertical, movement due to collapse, removal or displacement of underlying strata.
(a) Subsidence due to coal mining
Earliest excavations for coal were of the '*bell-pit*' type. Backfilling at the time would now be stable but would probably undergo additional *settlement* under a surcharge load. Subsequent deep excavations using '*stoop and room*', '*pillar and stall*', '*bord and pillar*' methods, wherein the coal was removed from headings leaving pillars to support the roof and were not backfilled, present serious subsidence hazards, as the pillars may still be slowly disintegrating and reaching a stage of collapse. In subsequent variants of the

method, the pillars themselves were worked, retreating towards the access shaft so that the workings subsided. From about 1850 onwards *'longwall'* methods of mining allowed of almost complete extraction of the coal with little danger of subsequent subsidence. Serious subsidence hazards also exist with the probable collapse of old shafts.

(b) Subsidence due to other causes
Subsidence may also occur due to collapse of strata following the underground mining of limestone (the Silurian Limestone around Dudley, Scottish Lower Carboniferous and Bath freestone at Box and Corsham in Wiltshire in Britain are examples), the mining of chalk and flint (dene holes and chalk wells), ancient open-cast and underground working of iron ore, the mining of oil shale, gypsum anhydrite and potash, metallic ores and associated gangue minerals, paving stones, sandstones, underground brine pumping and poorly backfilled areas worked for a variety of other reasons.

subsoil That part of the soil profile immediately below the *topsoil*.

substructure That part of any structure (including building, road, runway or earth work) which is below natural or artificial ground level, or in a bridge including piers and abutments (and wing walls) which support the superstructure.
Reference: CP 2004 : 1972.

subsurface map A geological or engineering geological map that shows the disposition of the rocks at some horizon below the actual ground surface. The horizon is normally one of engineering or geological significance, such as rockhead level, etc.

subsurface water All water below ground level.

suction lift The vertical distance that water can be lifted by suction (or vacuum) from source to inlet of the pump, taking into account any friction losses.

sulphate attack of concrete The condition whereby soluble sulphates in the ground react with Portland cement to form compounds (calcium sulphate and calcium sulpho-aluminate) which on crystallisation increase in volume and cause expansion and disintegration of the concrete. *Building Research Establishment Digest 250 (1981)* gives recommendations regarding minimum cement contents and water/cement ratios for concrete exposed to sulphate attack and the use of sulphate-resisting cement, depending on the degree of resistance required.

sulphate-reducing bacteria A group of bacteria found in most soils and natural waters, but active only in conditions of near neutrality and freedom from oxygen. They reduce sulphates in their environment, with the production of sulphides.
Reference: CP 1021 : 1979.

Sultan and Seed method of slope stability   A method for determining the stability of a sloping core in an earth dam using a sliding block analysis.
   Reference: 'Stability of sloping core earth dams', H. A. Sultan and H. B. Seed, *Jl Soil Mech. Found. Div., Proc. Am. Soc. Civil Engrs.*, July 1967.

sump   A depression formed in the bottom of an excavation, from which water can be pumped to keep the base free of water.

superficial deposits   Materials overlying the 'solid' or 'bedrock' which have resulted from erosional or human processes to shape the present land surface.

superposition   (1) The geological law that states that in an undisturbed sequence of strata, the younger will overlie the older. (2) The mathematical principle that states that the sum of any two solutions to a linear problem will itself be a solution. The principle is employed in *elastic analysis* when a complex pattern of loading is broken up into simple elements and the individual effects of each calculated separately. The total effect is taken to be the sum of the separate solutions.

supragelisol   The zone above the *permafrost* table, normally the *active layer*.

supraglacial tills   *See till.*

surface recycling   A process in which an asphalt pavement surface is heated in place, scarified, remixed, relaid and rolled. Asphalts, recycling agents, new asphalt hot mix, aggregates, or combinations of these, may be added to obtain desirable mixture characteristics. When new asphalt hot mix is added, the finished product may be used as the final surface. Otherwise, an asphalt surface course should be used. *See cold mix recycling; hot mix recycling.*
   Reference: *Civil Engineering*, January 1981.

surface run-off   Rainfall that flows to surface water sewers and streams, etc., over the ground surface.

surface tension   *See capillarity.*

surging   Methods of *well stimulation* which produce movement of water through a formation in order to flush out fine material and thereby increase its *permeability*. Various methods are employed, including use of surge blocks, varying water levels in the well by applying air pressure or intermittent use of the pump, and using chemicals to penetrate the borehole wall.

suspended floor slab   *See ground slab.*

S-wave   *See transverse wave.*

Swedish circle method of slope stability   A method of analysis developed by W. Fellenius in 1927 for the stability of a homogeneous earth slope subjected to steady seepage conditions. The mass above a trial failure surface was split into a number of vertical slices and the stability of each considered separately.

Swedish foil sampler A sampler in which soil is encased in metal foil as it enters the sampler, thereby eliminating sliding resistance between the sample and the inside wall of the sampler. Allows of sampling of up to about 25 m of soft cohesive soil in one operation.

Swedish weight penetrometer A penetration test equipment commonly used in Scandinavia and Finland to assess soil strength, and comprising a screw-shaped point which is loaded in steps, the strength of the soil being assessed from the amount and rate of penetration.

swelling index (C_s) The average slope of the unloading/reloading cycle in a semilogarithmic plot of effective pressure against *void ratio*, obtained from a *consolidation* test on a soil sample.

swelling pressure Certain rocks and soils, particularly those that contain clay minerals, have a tendency to swell when the load on them is removed. This causes a reduction in strength which leads, in the case of an underground opening, to a transfer of load to the lining of the excavation. This load is often termed *swelling pressure*.

swelling ratio The ratio between axial swelling (E_a) and radial swelling (E_r). When compacted expansive clays are allowed to swell, E_a/E_r decreases with time but reaches an equilibrium value termed the *equilibrium swelling ratio*, this being a function of the boundary loading conditions.

Syledis A portable survey precise range measurement positioning system similar in principle to the *mini ranger/trisponder* system. Digitally coded transmission and signal processing are used which allow low-frequency, long-pulse waves to be employed which tend to overcome the problems associated with phase comparison positioning systems. The range is between about two and four times the line of site (i.e. approximately 70–130 km), dependent on antenna model and amplifier unit.

T

tacheometer Basically a transit *theodolite* adapted for distance measurement by the provision of stadia hairs. By sighting on to a staff or stadia rod, the distance from station to instrument can be calculated. Published tacheometric tables allow of rapid determination of distance.

tacheometry A branch of angular surveying in which horizontal and vertical positions of points are determined purely from instrumental observations.

TACSS Acronym for Takenaka Aqua-reactive Chemical Soil-stabilisation System. A chemical one-shot grout system which does not gel until it reacts with water, developed by Takenaka Komuten Co Ltd, of Japan. It is designed to utilise the pore water in the soil as one of the two reactants. The chemical grout injected as the other reactant captures *groundwater* and reacts with it to form an insoluble and infusible gel.

Tagg method   A method of interpreting geophysical resistivity data obtained from a *Wenner configuration (see Figure E.1)*.

talik   A body of unfrozen soil within the *permafrost* profile.

Tams double-tube auger sampler   A sampler developed in 1951 for the recovery of undisturbed samples of silty sand and sandy silt without the use of a drilling fluid, and generally similar in operation to the *Denison sampler*.

Tams double-tube core barrel soil sampler   A sampler developed in the 1950s and generally similar in principle to the *Denison sampler*.

tamping roller   A compaction machine similar to a *sheepsfoot roller* but having a higher ratio of foot area to cylinder area (generally > 15 per cent) and more suitable to the compaction of wetter cohesive soils.

tandem vibrating roller   *See vibrating rollers.*

tar   The residue of distilled coal after removal of volatile constituents.

tare   The empty weight of a vehicle or receptacle.

Taylor's stability numbers for earth slopes   A method of analysis for the stability of soil slopes based on the friction angle using total stresses. Soils with a range of values of *angle of shearing resistance* can be considered but a constant cohesion *(c)* with depth is assumed. For a given angle of shearing resistance (ϕ) the critical height (H_c) of a slope is made equal to $N_s(c/\gamma)$, where N_s is a dimensionless stability number depending on the slope angle (β), the friction angle (ϕ) and a depth factor (n_d) which expresses the depth at which the clay overlies a material of significantly greater strength—a controlling factor in determining whether the slip surface is shallow or deep-seated. As total stresses are involved, the analysis gives only the immediate or 'end of construction' safe slope. *Figure T.1* shows relationships between N_s, β and ϕ.

Taywood pilemaster   Silent piling equipment for driving straight lines of sheet piles in cohesive ground; 200 tonne weight is applied to the head of the pile by use of the reaction of adjacent piles.

TDM   A chrome lignin product with a low viscosity used for grouting fine-grained soils to reduce *permeability* and increase strength.

tectonic, tectonism   Having regard to deformation, particularly on the large scale.

tectonic fabric   A rock fabric, consisting of preferred grain orien-

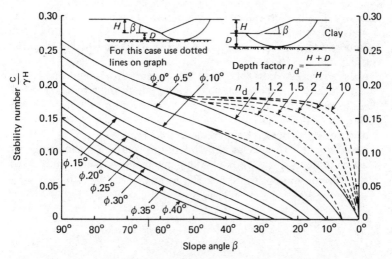

Figure T.1   Taylor's method of slope stability analysis

tations, schistosity, etc., which has been imposed by deformation subsequent to the initial formation of the rock.

tectonic map   A *geological map* that shows the disposition of tectonic units—i.e. areas of rock that have suffered similar histories of deformation.

Tellurometer   An electronic survey instrument designed to measure distances to a high order of accuracy by measuring the time taken to send and receive an electronic signal. The method involves measuring the time in millimicroseconds for radio waves to travel from a master unit, which takes the actual measurements, to a reflecting instrument and back again. Thus, if the velocity of radio waves is known, the slope distance between the two terminals can be accurately determined. *See also Autotape; Hydrodist.*

temperature log   A log of the temperature variation with depth obtained by lowering a recording thermometer or non-recording temperature probe down a borehole or well. The method has a number of uses in the oil and water well industries and in geotechnical surveying, including location of cement tops, gas entry and water channels, finding leaks in *casing*, lost circulation zones and measuring thermal gradients in carboniferous fills to help determine the possibility of spontaneous combustion occurring, etc.

temporary works   Defined in the British ICE Conditions of Contract as the temporary works of every kind required in or about the construction, completion and maintenance of the works.

241

tendon The bar, strand or wire in a *ground anchor* which takes the tensile force.

tensile strength The value of the *tensile stress* at which a material fractures or yields.

tensile stress A normal stress which causes elongation in the direction of the coordinate axis along which the stress is applied.

Tensiltarp A tough impermeable plastic sheeting manufactured by British Cellophane Ltd.

tensiometer A device for measuring *soil suction*.

tension cracks Cracks at the top of slip surfaces in cuttings or embankments due to the soil failing in tension. K. Terzaghi showed the maximum depth of crack Z_0, to be $(2c/\gamma)\sqrt{N_\phi}$, where $c =$ the soil cohesion; $\gamma =$ the bulk density of the soil; and

$$N_\phi = \tan^2 \left(45° + \frac{\phi}{2} \right) = flow\ value$$

N_ϕ is identical with the *coefficient of passive earth pressure*, and where the *angle of shearing resistance* is very small, N_ϕ approximates to unity and Z_0 becomes $2c/\gamma$.

tension pile A pile which is designed to resist an upward force.
Reference: CP 2004 : 1972.

terrain analysis The evaluation of an area of land by some criterion of usage—e.g. foundation conditions, road construction, etc. In geotechnics the technique has been adapted to reveal gross subsurface information on the basis of surface form, drainage, etc., and so is closely related to *photogeology*. The essence of the technique is to divide a landscape into individual elements, and to assign to each element some property of interest to the particular application, such as slope angle, ease of drainage, etc. The complete terrain can then be analysed element by element, and the results portrayed graphically. The technique has found most application in civil engineering evaluation in areas where detailed geological background information is not available.

terrain classification *See land system.*

terrain conductivity meter Portable instrument used for mapping small near-surface anomalous features, incorporating an inductive/electro-magnetic method of geophysical investigation. In the Geonics EM-31 meter, the transmitting and receiving coils are located at the ends of a 4 m long boom which can be carried by one man. The instrument is normally carried along a number of traverse lines and then recorded conductivities plotted to show up any anomalies. *See also ground probing radar; GSSI impulse radar system.*

terrain correction A correction applied to observed values of gravity during the calculation of the **Bouguer anomaly**. The correction allows for the presence of large masses of rock at higher elevations than the recording station—e.g. the mountain walls surrounding a station on the floor of a valley.

Terranier grouts Grouts of low molecular weight polyphenolic polymers which form insoluble gels of high strength when dissolved in water in the presence of an appropriate *catalyst*.

Terra-probe The Foster 'Terra-probe' method of soil densification comprises a vibratory pile driver and a probe consisting of a 0.75 m diameter, 13.5 m long open-ended steel pipe. The pipe is driven into the ground to the required depth of compaction, and withdrawn while vibrating vertically. The effect of both driving and vibrating the probe compacts the soil.

Terresearch (TSO) sea-bed sampler A system which can be lowered onto the sea-bed to allow cone penetration tests and thin-walled sampling to be undertaken at 1 m intervals for a total penetration of 10 m into the sea-bed.

tertiary creep *See creep.*

Terzaghi, K. Founder of the science of modern soil mechanics and first president of the International Society of Soil Mechanics and Foundation Engineering. Author of *Erdban Mechanik* and *Theoretical Soil Mechanics*, co-author with R. B. Peck of *Soil Mechanics in Engineering Practice*, and author of numerous technical papers and reports.

Terzaghi solution Karl Terzaghi presented a solution for the ultimate bearing capacity (q_u) for long footings in his *Theoretical Soil Mechanics* (1943), which was more general than had been given previously. Two cases were presented: (1) the case of general shear, where

$$q_u = C(N_c) + \gamma B(N_\gamma) + \gamma D_f(N_q)$$

and (2) the case of local shear where

$$q_u = C(\tfrac{2}{3}N_c') + \gamma B(N_\gamma') + \gamma D_f(N_q')$$

where N_c, N_γ and N_q = dimensionless bearing capacity coefficients *(Figure T.2)* dependent only on the angle of shearing resistance (ϕ) of the soil; B = the half-width of the footing; C = the soil cohesion; γ = the soil density; and D_f = the depth of footing below ground level.

Reference: *Theoretical Soil Mechanics*, K. Terzaghi, Wiley, New York, 1943.

test pile A pile to which a load is applied to determine the load–settlement characteristics of the pile and the surrounding ground.

Reference: CP 2004 : 1972.

thaw consolidation If a frozen soil is allowed to thaw, an excess pore pressure is generated which causes *consolidation* of the soil when it

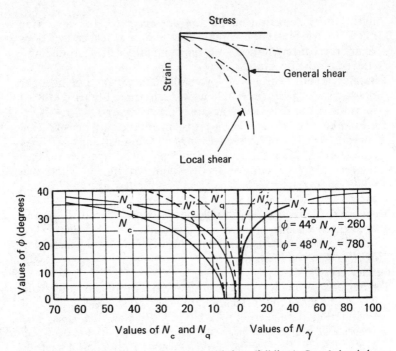

Figure T.2 Terzaghi solution: *Case 1:* general shear (full lines), *Case 2:* local shear (dotted lines)

dissipates. This consolidation takes place without overall change in the *effective stress,* and is due to the disruption and subsequent settlement of the soil fabric as a result of the formation of ice in the soil pores. A theory of thaw consolidation has recently been evolved which allows of the prediction of the pore pressures and the degree of settlement. This finds applications in ground engineering in *permafrost* areas—in particular to the analysis of the stability of thawing slopes, and the prediction of settlements beneath heated buildings and pipelines.

Theis method   A method with which the coefficients of *transmissibility (T)* and *storage (S)* of an *aquifer* can be calculated using non-equilibrium well formulae. It requires the measurement of *drawdown* in one or more observation holes when pumping from a *well* sunk into the aquifer and matching the drawdown plots with 'type curves'. From the information so obtained:

$$\text{the coefficient of transmissibility } (T) = \frac{QW(u)}{4\pi(Z - Z_0)}$$

244

and

$$\text{the coefficient of storage } (S) = \frac{4t\,Tu}{r^2}$$

where Q = constant pumping rate: $(Z - Z_0)$ = drawdown at time t; r = the distance from the well to the observation hole; $W(u)$ = the well function, where $u = (r^2 S)/(4tT)$. *See also* **Cooper–Jacob** *analysis; Prickett analysis.*

Reference: 'The relation between the lowering of the piezometric surface and the rate and duration of discharge of a well using a groundwater storage', C. V. Theis, *Trans. Am. Geophys. Union*, **16**, 1935.

theodolite A surveying instrument for measurement of horizontal and vertical angles to a high order of accuracy. Basically it comprises a telescope with cross hairs enabling precise pointing of the instrument, with a magnification from about × 20 to × 40 and a field of view of about 25–30 m at 1 km. Spirit levels allow the horizontal and vertical axes to be defined and measurement is by reference to optical glass graduated circles.

theodolite station The accurately surveyed-in location of an observation point over which a *theodolite* is centred.

thermal conductivity of soil The rate at which heat flows through soil. A knowledge of this is sometimes required in the calculation of heat flow to and from buried objects such as cables, ground coils for heat pumps, and pipes conveying warm liquids and gases. The traditional method of measurement involves applying a constant power input to a spherical or cylindrical body placed in the soil and recording the temperature rise. Two more rapid and cheaper methods are the *soil constitution method* and the *transient needle method*. The former involves testing an undisturbed sample in the laboratory to determine its dry density moisture content and clay fraction, which are used in a nomogram developed from formulae by Gemant to give thermal conductivity. The latter method involves the *in situ* placement of a long thin heater and measuring temperature increase with time by either a resistance thermometer or a thermocouple.

Reference: 'The thermal conductivity of soil', A. Gemant, *Jl Appl. Phys.*, **750**, p. 21, 1950.

thermal decay–time log *See neutron-lifetime log.*

thermal infra-red That part of the *electromagnetic spectrum* having wavelengths between 3 μm and 1000 μm.

thermistor A resistance thermometer probe wherein the voltage drop across a platinum wire resistance is a function of the temperature effect on that *resistance*.

thermography *T hermal infra-red* imagery, being a record of emitted thermal energy, and analogous to photography, being a record of reflected energy.

thixotropy The property of some gels to liquefy when shaken but to reset quickly on standing.

three-arm protractor An instrument similar to a *station pointer* (*Figure T.3*).

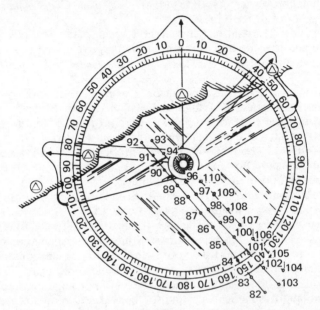

Figure T.3 Three-arm protractor

Three-D Log A trade name of the Birdwell Division of the Seismograph Service Corporation for a three-dimensional velocity log which displays the seismic or acoustic wavetrain received a short distance away from a sonic wave transmitter.

three-point problem A classic geometric problem which arises in the interpretation of geological data. The problem is to find the attitude of a planar bed, knowing the elevations of three points which are not colinear. The method is to find the line of slope between the highest and the lowest point, and by division of this line pro rata to find the point on it which has the same elevation as the remaining third point. Thus, the first *strike line* can be drawn and the others then drawn parallel to it, the spacing being obtained from the drop on the initial line.

thrust boring, or pipe jacking Thrust boring basically consists of transmitting a horizontal force from a vertical ground surface by

246

means of large capacity hydraulic jacks which thrust a concrete or steel pipe train forward at the same time. The method is very competitive in soft ground for pipe diameters up to 1.5 m for lengths around 100 m and up to 2 m or more for short distances— e.g. under railways or canals.

Reference: 'Pipe jacking basics', *Civil Engineering*, September 1979.

tidal range The difference in height between one high water and the preceding or following low water.

tide The periodic vertical movement of water generated primarily by the gravitational pull of the Sun and the Moon. Non-periodic movement is caused by factors such as wind, atmospheric pressure, precipitation and evaporation. Tidal range is affected by geographical considerations and can vary from as little as a few millimetres in the open ocean to, for example, in excess of 15 m in the English Severn Estuary.

tide gauge A calibrated vertical board set in a river or harbour to allow of measurement of tidal water level relative to a datum. *Automatic tide gauge* recorders allow water levels to be continuously recorded on a revolving-pen chart recorder. Permanent gauges are normally set to Chart Datum for the standard port nearest to the particular locality. In the United Kingdom these values relative to **Ordnance Datum** Newlyn can be obtained from Admiralty Tide Tables.

tide pole or staff A simple graduated pole set vertically in the water to allow changing level of the sea surface to be read on it.

till The general name given to the sediment deposited directly by a glacier. It is characteristically an unsorted aggregate of those rock types over which the ice has moved, and may contain any size of particle. A distinction is made between *supraglacial tills* (sometimes known as **ablation** tills), which are deposited on the ice surface, and *subglacial tills*, which are deposited beneath the glacier. The former are divided into *melt-out till*, which is formed when interstitial ice is removed from frozen debris without any other disturbance, and *flow till*, which is produced when melt-out debris moves as a *mudflow* down the glacier surface. Subglacial tills are divided into *lodgement till*, which is produced when material held in the glacier sole is frictionally retarded against the bedrock, and *ice-side till*, which forms when material melts out in an unconfined space beneath the glacier.

time factor (T_v)
$$T_v = \frac{tC_v}{d^2}$$

where t = the elapsed time since the application of a change in total normal stress; C_v = the *coefficient of consolidation*; and d = the length of the *drainage path*.

TNO method of pile testing System developed in the mid-1960s by the Netherlands Technical Organisation whereby the tip of the pile is struck by a 1.5 kg fibre hammer and the time taken for the resultant reflected compressive shock wave to return to the pile head is detected by an accelerometer. This is hand held on the pile head which does not require special preparation. The wave is transmitted to, and integrated by, a signal processor for displaying vertical displacement against time in milliseconds on the horizontal scale of an adjacent oscilloscope. A built-in instant print camera can record the tests for future reference. The method can be used to detect and assess pile deformities and can be used as a form of quality control, 50–100 piles being tested per day usually starting at least one week after casting.

Reference: *New Civil Engineer*, 11 June 1981.

tool pusher An American oil industry term for the foreman in charge of a drilling rig.

toppling A form of instability which occurs in rock slopes where discontinuities behind the face are steeply inclined towards the face and allow individual pieces of rock to fall under gravitational forces. This action is aided by ice or water pressure in the joints (*Figure T.4*). See *discontinuity; joint*.

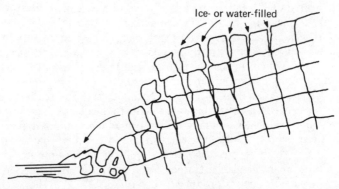

Ice- or water-filled

Figure T.4 Toppling failure

topsoil The thin outer skin of the Earth's crust which has become capable of supporting plant growth, comprising mixtures of fine-grained rock particles and organic material resulting from the physical and chemical weathering of rocks and other agencies.

Toran A medium-range electronic positioning system using phase-comparison methods similar to *Raydist, Lorac* and *Decca*.

total core recovery (TCR) A parameter used in the description of rock core recovered from a borehole, and defined as the summed length of all the pieces of recovered core expressed as a percentage of the length drilled.

total energy of a fluid The sum of a fluid's potential, kinetic and pressure energies:

$$potential\ energy = WH$$

where W = unit weight of fluid and H = the maximum height through which the fluid can fall.

$$kinetic\ energy = \frac{Wv^2}{2g}$$

where v is the velocity at which the fluid is moving.

$$pressure\ energy = P = WH$$

where P is the intensity of pressure at depth H below the free surface of a fluid of unit weight W. Provided that a continuous hydraulic connection exists between all the particles comprising the mass of a fluid, the total energy of each particle is the same *(Bernoulli's theorem)*.

total sound level The addition of the (usually) many sounds occurring at a given location, e.g. a construction site. Sounds from different sources combine approximately in accordance with the following table:

Difference between two sound levels dB(A)	Addition to higher level dB(A)
0–1	3
2–3	2
4–9	1
>9	0

*See also **sound; noise; sound level; equivalent continuous sound level; sound attenutation; decibel; reflected sound.***

total stress The force per unit area, or pressure, transmitted across a plane in a mass of soil. An increase in total stress in a fully saturated soil is initially carried entirely by the pore water. This causes a pressure gradient in the pore water, which flows towards a free-draining boundary of the soil layer, until the force is carried by the solid particles in the soil structure and equilibrium is restored. *See **effective stress.***

total unit weight   The ratio between the total weight of a material and its total volume.

toughness index   *See Atterberg limits and soil consistency.*

towed vibrating rollers   *See vibrating rollers.*

tractor and scraper   A type of plant that can excavate, transport and then unload and spread the material to the required thickness under its own power. Machines range from about $10\,m^3$ to $25\,m^3$ capacity and rubber-tyred versions can travel quickly with economic haul loads of up to about 2000 m. The plant is often assisted by track-laying tractors where surfaces are soft and slippery.

traditional strip foundations   *See strip foundation.*

transducer   A device for converting energy from one form to another. Various types exist, including those making use of wire resistance *strain gauges*, vibrating wire gauges, semiconductor strain gauges and piezoelectric crystals; displacement transducers used in foundation instrumentation, including linear variable differential transformer and potentiometric displacement types; and mechanical dial gauges and photoelastic types.

transient-needle method   A method used for the rapid in situ measurement of *thermal conductivity of soil*. *See also soil constitution method.*

transit line   An imaginary line extending through two *transit markers*. Vessel positioning can be accomplished by an observer on the vessel sighting pairs of markers set on two intersecting transit lines.

transit markers   Transit markers can be either significant existing features on shore or along a shoreline, such as church spires, tall chimneys, etc., or purpose-made targets, usually of wooden construction, and given identification by the use of various coloured shapes—e.g. red diamond, white circle. The imaginary line extending through two such markers is termed a *transit line*.

transit position fixing system   The US Navy system of position fixing using five artificial satellites in circular polar orbits between 700 and 1100 km high with periods of about 100 min. Signals are emitted and received at tracking stations around the world which compute present and future orbital parameters and transmit these back to the satellites, which, in turn, continuously retransmit the information. The information is used by the receiving ship with measurement of the satellite's Doppler frequency shift to allow its position to be determined.

translational slide   A form of instability that can occur as a result of planes of weakness in soil and rock masses, such as bedding planes, faults, cracks and fissures running approximately parallel to the ground surface *(see Figure T.5)*. Gravitational forces acting on the soil or rock above such planes result in a translational downhill

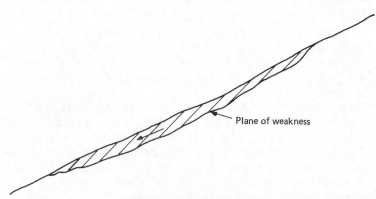

Figure T.5   Translational slide

movement of the material. Silt or sand lenses running roughly parallel to the ground surface may also become potential weakness planes where excess pore pressures due to heavy rain can cause a high reduction in strength of the lense material.

transmissibility   The *coefficient of permeability (or hydraulic conductivity)* multiplied by the thickness of an aquifer.

transmissivity   The rate at which water at a particular kinematic viscosity is transmitted through an *aquifer* of unit width, under a unit *hydraulic gradient*. This term replaces the term *transmissibility*.

transpiration   The process whereby moisture in plants and trees is returned to the atmosphere. During long dry summer periods, transpiration losses from trees and hedges can be very high, causing severe desiccation around their root systems with resultant damage to foundations where these are constructed at shallow depth in soils of medium to high *plasticity*.

transverse wave   An elastic body wave such as a *shear or S-wave*, in which the motion of the particles vibrate in a plane at right angles to the direction of wave propagation. The velocity of the waves is

$$V_S = \sqrt{\frac{E}{\gamma} \cdot \frac{1}{2(1+\mu)}} = \sqrt{\frac{G}{\gamma}}$$

where E = Young's modulus of elasticity; γ = the mass density = density/g; μ = Poisson's ratio; g = gravitational acceleration; and G = the modulus of shear. *See also **P-wave, compressional wave**.*

Reference: *Geophysical Prospecting for Oil*, 3rd edn, Milton B. Dobrin, McGraw-Hill, 1976.

travel-time curve   In *refraction seismic prospecting*, a graph of the time taken for the first arrival *seismic wave* to travel from the source to the receiver plotted against the horizontal distance from source to receiver is termed a *travel-time curve*. For a uniform substratum

251

the 'curve' will be a straight line whose slope is the reciprocal of the *seismic velocity*. If the substratum is layered, breaks of slope will occur in the travel-time curve, and from these may be deduced the depths to the interfaces. If the interfaces are irregular, the travel-time curve also becomes irregular, and in this case the method of *delay times* may be preferred.

Tremie injection or placement A technique of placing concrete under water by allowing it to gravity feed from the hopper through a pipe, the bottom (discharge) of which is kept below the concrete level to avoid segregation occurring and to avoid dilution by the water. The injection process is similar in that cement grout is led down the grout pipes directly into the cavities to be filled.

trench-fill foundations *See strip foundations.*

trend surfaces (fitting of) *See multivariate.*

triangulation The traditional method of 'carrying' position across the Earth's surface by a network of geometrical figures extended by angular measurement from a baseline of known length and astronomical observations for azimuth.

triaxial compression machine Laboratory equipment for testing a generally cylindrical soil specimen, which permits of independent control of the three principal stresses and allows of examination of generalised states of stress *(Figure T.6)*. The specimen is sealed in a rubber membrane and enclosed in a cell subjected to water pressure. An axial load is applied through a ram to control the *deviator stress* (the *major principal stress*, σ_1). The *intermediate principal stress*, σ_2 and the *minor principal stresses*, σ_3 are each equal to the applied cell pressure. Three basic types of shear strength test are commonly carried out—i.e. *undrained tests*, *consolidated undrained tests* and *drained tests (see shear strength)*. From these tests values of *cohesion* and *angle of shearing resistance* can be obtained in terms of either *total stress* or *effective stress* for use is appropriate stability problems. The equipment can also be used to study other soil properties, including those for *consolidation, permeability, pore pressure*, and K_0 (earth pressure at rest). Detailed information on triaxial testing procedures is given in *The Measurement of Soil Properties in the Triaxial Test*, by A. W. Bishop and D. J. Henkel, Arnold, London, 1957.

triaxial compression test *See shear strength; triaxial compression machine.*

trilateration A modern method of 'carrying' position across the Earth's surface by a network of geometrical figures extended by measurement of their sides using electronic distance measuring devices such as the *tellurometer*.

trisponder A marine survey instrument for distance measurement comprising a shipborne station (Mobile) housing the Distance

252

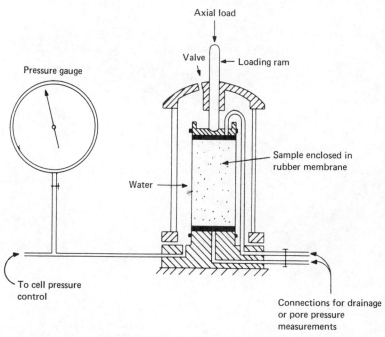

Axial load

Valve

Loading ram

Pressure gauge

Sample enclosed in
rubber membrane

Water

To cell pressure
control

Connections for drainage
or pore pressure
measurements

*Figure T.*6   Triaxial compression machine

Measuring Unit (DMU) and a Base Unit, and up to four
transponder stations (Remotes), installed on shore or in buoys.

true specific gravity   A *rock mechanics* test whereby a rock specimen
is pulverised to overcome the effects of closed, non-intersecting
pores in determining true unit volume.

true or total porosity   A *rock mechanics* test, being the ratio between
the *void ratio (e)* of a specimen and the value $e + 1$, e being obtained
from the expression $(G/G_A) - 1$, where $G =$ the *specific gravity*, and
$G_A =$ the *apparent dry specific gravity*.

tsunamis   Waves caused by seismic disturbances at the bottom of the
sea. Tsunamis travel considerable distances and can create heavy
damage by inundating coastal areas.

tube à manchette   A double *packer* grouting device invented by E.
Ischy in 1933 for use in alluvium *(Figure T.7)*.

tube well dewatering   Tube well dewatering consists of placing a
temporary *casing* in the ground by some form of boring technique
and then inserting a perforated well liner, the annular space
between liner and casing being filled with a filter medium (***gravel
pack***) to allow of the flow of clean water through the well screen.
The temporary casing is withdrawn and an electric ***submersible***

253

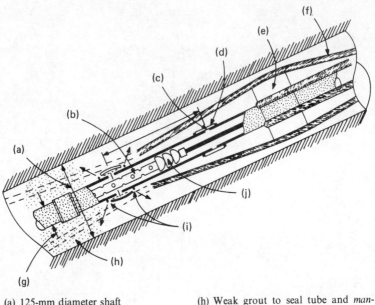

(a) 125-mm diameter shaft
(b) Grouting pipe
(c) Grouting orifice
(d) Rubber *manchette*
(e) Plastic spacer
(f) Strand
(g) 50-mm diameter *tube à manchette*

(h) Weak grout to seal tube and *man-chette* into hole
(i) Grouting pressure distends rubber *manchette* and forces grout through sealing grout
(j) Grouting head with double packer top and bottom

Figure T.7 Detail of *tube à manchette* for pressure grouting control

pump is lowered down the well. The amount of *drawdown* (head) is limited only by the capacity of the pump. Single- or multiple-well arrays may be used, depending on requirements. The system can be economic where more than one stage of *wellpointing* is necessary but cannot be used, because of space limitations or other reasons.

Tubex pile The Tubex pile is similar in principle to a *Fundex pile*, except that the drive tube is left in position. *See also* *pile foundations*.

turbidity The cloudiness caused by suspensions of clay, silt, finely divided organic matter, plankton and other microscopic organisms in water.

turbulent flow The motion of fluid which occurs above the higher *critical velocity* and sometimes down as low as the lower critical velocity, as opposed to *laminar* or streamline *flow*.

U

Udden–Wentworth scale *See Wentworth scale.*

ultimate bearing capacity *See bearing capacity.*

ultrasonic pulse velocity A non-destructive test used in *rock mechanics*. Both compression (longitudinal or *P-wave*) and shear (transverse or *S-wave*) may be measured to provide an indication of the soundness of the rock.

unconfined aquifer *See groundwater (unconfined).*

unconfined compression test Initially a field test wherein a cylindrical cohesive soil specimen, approximately 75 mm long by 38 mm diameter, was subjected to an axial compressive force, a simple spring-loaded hand-operated portable machine being used to determine the *compressive strength* of the soil. Undrained triaxial compression tests carried out with zero lateral pressure are essentially unconfined compression tests and for saturated clay soils with zero *angle of shearing resistance*, the value of compressive strength obtained in either test is about equal to twice the *cohesion* or *shear strength*. The field or laboratory test is of limited value if the clay contains a significant proportion of silt or sand and, hence, has an *angle of shearing resistance* (ϕ). The test is also carried out to determine the confined compressive strength of rock specimens, with large compression testing machines.

unconfined groundwater A body of *groundwater* where the upper surface is the free water surface, and, hence, is at atmospheric pressure.

unconsolidated An imprecise term used in geology to mean loose or uncemented. It has no direct relationship to the state of *consolidation* of a soil in the sense of being *normally* or *overconsolidated*.

underconsolidated Under (or partial) consolidation results when insufficient time is available for full dissipation of pore pressures set up by overlying deposits and the soil is therefore not subjected to the maximum potential *effective stress*. Theoretically a clay can only become fully consolidated if it is formed by an infinitely slow deposition of soil particles such that the pore pressures dissipate as the thickness of the stratum builds up. Gibson (1958, *Géotechnique* **8**, pp. 171–182) showed that for a stratum of thickness h built up at a uniform rate of deposition r, the ratio C_v/rh would need to be greater than about 10 in order that the clay reached approximately 95 per cent of the maximum possible consolidation (a practical limit), C_v being the *coefficient of consolidation* and typically 1 m^2/year. Most clays are deposited sufficiently slowly to satisfy this condition, but rates of deposition in modern deltas such as the Orinoco Delta in the Caribbean and the Mississippi Delta can be

255

very rapid (100–1000 m/years or more), which leads to significant underconsolidation and consequent great thicknesses of very soft clays—a problem in the design and construction of offshore gas/oil exploration structures.

underpinning The transfer of loads from the original foundations of a structure to new deeper *strip, pad* or *pile foundations*. It is usually carried out when damage to building superstructures (especially houses) caused by trees, soil erosion and other phenomena is being rectified.

under-reamed piles *See bored and cast-in-situ concrete piles.*

undisturbed sample A sample taken from the ground in such a way that it suffers minimal disturbance from the condition pre-existing before sampling takes place. Sampling equipment varies, depending on the type, strength and fabric of the soil or rock. Removal of the material from its natural environment must cause some disturbance.

undrained test *See shear strength; triaxial compression machine.*

uniaxial compression test A *rock mechanics* test used to determine the uniaxial *compressive strength* of rock by compressing a cylindrical specimen in a compression machine.

uniaxial tensile test A *rock mechanics* test for the direct measurement of tensile strength of rock specimens. The test is not common, as results are usually rather scattered.

unified soil classification system This system evolved from the *airfield classification system* after modifications by A. Casagrande of Harvard University and the US Bureau of Reclamation and the Corps of Engineers. It classified soil into three general groups—i.e. coarse-grained, fine-grained and highly organic, with a further fifteen categories depending on gradation of the coarse-particled soils and *plasticity* of the fine-grained soils.

uniflote A marine craft used for drilling platforms, etc., which can be built up to the desired size from individual units.

uniform taper piles *See Raymond piling systems.*

uniformity coefficient The ratio between the D_{60} size on a grading curve obtained from a sieve analysis and the D_{10} size (the *effective grain size*), the D_{60} size being that size where 60 per cent of the material is finer and 40 per cent coarser. The ratio D_{60}/D_{10} gives a measure of the soil grading, a well-graded soil having a fairly flat grading curve and a uniformly graded soil (i.e. approaching a single particle size) having a steep curve.

unit boiler horsepower *See horsepower.*

unlimited flow strain, or unlimited flow deformation In relation to *liquefaction*, flow strains or deformations that commence following liquefaction and continue unabated under undrained constant *total stress* conditions. These deformations are accompanied by permanent loss of shear strength. This behaviour has been called 'true liquefaction' by some engineers in the past.

Reference: 'Definition of terms related to liquefaction', *ASCE Jl Geotech. Eng. Div.*, **104**, Pt GT9, September 1978.

USSR Building Code (1955)—values of limiting differential settlement The amount of differential settlement that a structure can undergo before causing distress to the building. *See Table 14; see also Bjerrum's danger limits for distortion of structures.*

Reference: 'Maximum allowable non-uniform settlement of structures', D. E. Polshin and R. A. Tokar, *Proc. 4th Int. Conf. on Soil Mech. and Found. Eng.*, **1**, pp. 402–405, London, 1957.

USSR mantle cone *See penetrometer (apparatus).*

V

vacuum well A *well* which is sealed at the surface to prevent the entry of air and to which a vacuum is applied to increase the yield from deep *aquifers.*

vadose A term applied to water occurring between ground level and the *water table.*

valley bulging Deformation of soft strata in the floor of a valley. Valley bulging has been observed mainly in valleys in southern Britain and is associated with cold climate processes at the end of the Pleistocene. It is believed that stress release or valley side movements contribute to the production of valley bulges, but the exact mechanism is unclear. Valley bulging is frequently associated with the *cambering* of cap rocks on the valley sides.

Van der Waals force An attractive force existing between atoms and molecules of a substance due to sympathetic movement of electrons in adjacent atoms and molecules.

vane test *See laboratory vane test; field vane test.*

Van Veen grab *See grab sampler.*

variance In statistics, the square of the standard deviation. The variance of a distribution is thus a measure of its spread. It is calculated by averaging the squares of the differences between the data points and the mean of the distribution.

varved clay Clay containing alternating thin layers of sand, silt and clay derived by seasonal melt water into fresh-water lakes at the close of the Ice Age.

velocity head The equivalent head or kinetic head of a fluid, due to its velocity. If water of weight W is allowed to fall freely through a height H under gravity, the potential energy WH is converted to a kinetic energy equal to $Wv^2/2g$ where g is the gravitational constant, from which $H = v^2/2g$. The form $v^2/2g$ is known as the velocity head. *See also total energy of a fluid.*

vibrated shaft Franki *See Franki piling systems.*

vibrating crushing test A test wherein an oven-dried sample of chalk is crushed and sieved to provide an *ad hoc* method of assessing the susceptibility of chalk to crushing and optimum method of handling chalk fill.

Reference: 'Étude par vibrovoyage de l'aptitude des craies au compactage', R. Struillou, *Bull. liaison Labo P et Ch.*, Spécial V, October 1973.

vibrating plate compactor A machine used for compaction of granular soils, basically comprising a steel base-plate with upturned edges, supporting either a fixed or pivoted engine-driven eccentric mechanism.

vibrating plate tamper Similar to a *vibrating plate compactor* but having a system of springs attached to the plate which is activated by a hollow piston within the main engine housing.

vibrating rollers Compaction machines similar to *smooth-wheeled rollers* but with the addition of an engine-driven vibrating unit that operates at pulse rates ranging from 1100 to 2000 pulses per minute. They are used mainly for compacting granular materials such as sand, gravel and rock fill. Their effectiveness depends on the correct choice of weight and vibration characteristics and may be categorised as follows:

(a) *Single-roll pedestrian-controlled rollers,* having a choice of petrol or diesel engine and a clear side to allow of compaction against obstructions such as kerbs.

(b) *Double-roll pedestrian-controlled rollers,* having vibration applied to both rolls.

(c) *Tandem vibrating rollers,* which normally have a vibrating smooth rear roll and a non-vibrating deadweight smooth front steering roll.

(d) *Double vibrating rollers,* which feature all-roll drive and vibration.

(e) *Towed vibrating rollers,* which have large roll diameters to reduce rolling resistance.

(f) *Self-propelled vibrating rollers,* which have a steerable vibrating front roll and a choice of rear rolls with traction tread tyres, smooth tyres or smooth steel rolls, depending on the application.

vibrating-string-type gravimeter *See gravimeter.*

vibro-corer A marine-bed sampler consisting of a barrel ranging from about 2 m to 12 m in length, provided with a cutter and a core catcher, which is driven into the sea-bed by vibrating motors attached to the top of the barrel, the whole mounted in a frame.

Vibro-corers may be driven by either compressed air, hydraulic or electrical means.

vibro-driver/extractor A vibratory hammer which can be used for installing or extracting caisson casings, heavy H-beam or bearing pile sections, concrete and long sheet steel piles in difficult ground conditions. Available in a range which includes machines capable of a dynamic driving force of over 1 MN and working line forces for extraction in excess of 400 kN.

vibro-flotation, or vibro-compaction A mechanical method of soil densification and of improving the bearing capacity of loose or soft ground, whereby an eccentrically vibrated cylindrical steel probe is jetted into the ground and the holes thus formed are filled with graded granular backfill as the mandrel is withdrawn to form compact stone columns tightly interlocked with the surrounding soil. *See also* ***vibro-replacement***.

vibro pile A pile formed by driving a steel tube with a conical cast-iron shoe into the ground with a 2030–4060 kg hammer. At the required depth, reinforcement steel and concrete are placed in the tube, which is then withdrawn by use of an extraction hammer, the movement of which compacts the concrete. *See also* ***delta pile; pile foundations***.

vibro-replacement Similar in principle to the ***vibro-flotation, or vibro-compaction*** method of soil densification, except that penetration of the mandrel is not assisted by jetting.

vibroseis A technique of seismic exploration in which a signal of continuously varying frequency is fed into the ground via a vibrating weight. Reflections from geological horizons are identified by statistical processing of the returns from several stations and correlating them with the waveform from the source. The advantages of vibroseis lie in the ability to tailor the frequency envelope to the particular characteristics of the geological horizons involved, and also in the low energy input used, which enables investigations to be conducted in urban areas.

virgin compression *Compression* of a soil under a load higher than has been experienced in the past, leading to plastic deformation of the soil.

virgin compression line The path of ***void ratio*** against ***effective stress*** taken by a soil undergoing ***virgin compression***. This path is sensibly straight for many soils if stress is plotted to a logarithmic scale.

void ratio The ratio between the volume of voids and the volume of solids in a material.

voids The volume (spaces or voids) of a material occupied by air and/or water.

volcanic water Juvenile water from lava flows and volcanic eruptions.

volumetric coefficient (VC) The ratio between the actual volume of a particle and that of a sphere in which it can just be enclosed:

$$VC = \frac{6v}{\pi a^3}$$

where v = the volume of the particle, and a = the largest dimension of the particle.

Shape of particle	VC	Shape of particle	VC
Sphere	1	Angular gravel	0.22
Cube	0.37	Plates	0.07
Tetrahedron	0.22	Needles	0.01
Round gravel	0.34		

vortex velocity anemometer *See anemometers.*

W

waisting of piles *See necking, or waisting, of piles.*

walings Horizontal beams supporting a cofferdam wall or poling boards or runners against external pressure.
Reference: CP 2004 : 1972.

wall piece A vertical member (usually timber) placed in direct contact with a wall to distribute the thrust from a *shore* or shores.
Reference: CP 2004 : 1972.

wall strength of discontinuity One of the ten parameters selected to describe discontinuities in rock masses, being the equivalent compressive strength of the adjacent rock walls of a *discontinuity*, which may be lower than the rock block strength owing to weathering or alteration of the walls; an important component of shear strength if rock walls are in contact.
Reference: International Society for Rock Mechanics. Commission on Standardisation of Laboratory and Field Tests.

Warlam piezometer A type of pneumatic piezometer.
Reference: 'Measurement of hydrostatic uplift pressure on spillway weir with air piezometers', A. A. Warlam and E. W. Thomas, *Instruments and Apparatus for Soil and Rock Mechanics*, ASTM, STP 392, pp. 143–151, 1965.

warp Laminated sandy and silty clay formed by periodic flooding of low-lying areas.

wash boring A method of boring wherein the hole is advanced by a combination of chopping and jetting to break up the soil and rock into small fragments (cuttings) which are flushed away by the drilling fluid. Allowing the cuttings to settle out in a sump provides 'wash samples' suitable for general identification of the strata being drilled.

water–cement ratio The ratio between the amount of water and the volume of cement for a given concrete mix design.

water displacement method for field density determinations *See field density tests.*

water replacement method for coarse material for field density determinations *See field density tests.*

water movement tracers Indicators added to water to trace its movement—e.g. in problems associated with sea disposal of effluent. These include artificial or added tracers such as dyes and radioactive isotopes, or measurement of natural inclusions such as salts (*salinity*) and bacteria. *See also* **radiotracers**.

water of saturation The total amount of water that can be absorbed by water-bearing materials without dilation.

water retentivity The ability of a grout to retain water against a vacuum. It is also a measure of the grout's *cohesion* and its ability to resist dilution in underwater work.

water table The level at which water stands in a *well* or *borehole* penetrating an unconfined water body, where the pressure is atmospheric.

water table (cone of depression) A cone of depression in the *water table* caused by drawing down the water level in an unconfined *aquifer* when pumping from a well. *See Figure D.3; cone of water-table depression.*

wave gauge A marine instrument for measuring the height and period of sea waves.

wearing course *See pavement design.*

wedge failure The bodily forward movement of a wedge of rock or clay in a cliff face, on two or three well-defined *joint* planes which intersect behind the face *(See Figure W.1)*.

weephole A hole provided through a retaining wall or other structure to allow of the discharge of water from the retained material and prevent the build-up behind the wall of excess *hydrostatic stress* or pressure.

weight-in-water method for field density determinations *See field density tests.*

weight sounding test (WST) The Swedish weight penetrometer is a device for measuring the *in situ* shear strength of soils, consisting of a screw-shaped point which can be loaded through rods by a series of weights. It can be used as a static penetrometer in soft soils (penetration resistance less than 1 kN) or rotated in stronger soils, when the number of rotations required to give a particular settlement is related to shear strength.

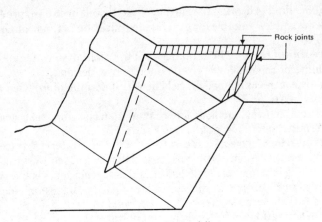

Figure W.1 Wedge failure

well   A borehole or pit sunk for the purpose of extracting water, gas or oil from the ground.

well efficiency   The ratio between the actual *specific capacity of a well* and the theoretical specific capacity. *See* Chapter 13, 'Well efficiency and maintenance', *Water Well Technology*, McGraw-Hill, for discussion on optimisation of well and pump efficiency, well corrosion and incrustation, *well stimulation*, etc.

well logging   The technique of lowering sensing devices down a *well* or *borehole* to determine the characteristics of the strata penetrated, the distribution of the contained fluids and, to some extent, the construction of the hole. Basic types are those which record electrical, acoustic, nuclear and miscellaneous measurements. The versatility of acoustic and nuclear logging equipment enables acoustic gamma ray, neutron, caliper and collar logs to be simultaneously obtained as routine in cased and uncased wells.

wellpoint   Essentially a shallow *well* comprising a small well screen of about 50 mm diameter by 0.5–1.0 m long, connected to a length of riser pipe which can be installed by jetting with a high-pressure water-jetting pump. Wellpoints are normally closely spaced (1–3 m) to enclose the excavation to be dewatered and connected to a common suction-header main and thence to a high-vacuum suction pump. *Groundwater* lowering is limited by the practical suction lift, normally about 5 m, in granular soils down to coarse silt to fine sand size. When deeper excavations below the water table are required, further stages of wellpoints, header mains and pumps must be used, each stage lowering the water table by about 5 m. Plastic disposable or 'throw-away' wellpoints are often used, especially where aggressive groundwaters exist. *See also* **eductor (ejector) wellpoint system**.

wellpointing A technique for lowering the *groundwater* level over a defined area by simultaneously pumping from a number of *wellpoints* inserted below the water table. Each wellpoint, a specially designed screened tube, is installed by jetting using water from a high-pressure pump.

well rehabilitation *See well stimulation*.

well screen A screen used in a *well* to prevent entry, during pumping, of the surrounding soil. The size of perforations in the screen is designed to allow initial entry of fines (silt and fine sand), so that the surrounding natural granular deposits and/or *gravel pack* build up into a filter which subsequently allows only clean water to be pumped.

well stimulation Well stimulation is known as *well rehabilitation* in the water-well industry and comprises techniques for reducing the effects of corrosion and incrustation of a well. It is defined by Koenig as the treatment of a well by mechanical, chemical or other means for the purpose of reducing or removing an underground resistance to flow, and includes *surging, jetting, acidising, hydraulic fracturing* and *shooting, or blasting*.

Reference: 'Economic aspects of water well stimulation', L. Koenig, *AWWA Jl*, **52**, No. 5, pp. 631–637, 1960.

Wenner configuration An electrode array *(Figure E.1)* used in *electrical resistivity* surveying, comprising four in-line equally spaced electrodes. An electric current is applied to the outer two electrodes and the induced potential is measured across the inner two. An *apparent resistivity* $= 2\pi aR$ is obtained, where a is the electrode spacing and R is the measured *resistance*.

Wentworth scale A scale of grain size for soils:

Description	Size (mm)	Description	Size (mm)
Boulder	>256	Medium sand	$\frac{1}{2}$ $\frac{1}{4}$
Cobble	256–64	Fine sand	$\frac{1}{4}$ $\frac{1}{8}$
Pebble	64–4	Very fine sand	$\frac{1}{8}$ $\frac{1}{16}$
Granule	4–2	Silt	$\frac{1}{16}$ $\frac{1}{256}$
Very coarse sand	2–1	Clay	$<\frac{1}{256}$
Coarse sand	$1-\frac{1}{2}$		

WES cell An earth pressure cell developed by the Waterways Experiment Station of the US Army Corps of Engineers generally similar to the *Carlson stress meter*, but the deflection of the diaphragm is recorded by means of *strain gauges*.

WES mobility cone penetrometer test A test developed by the Waterways Experiment Station of the US Army Corps of Engineers for off-road vehicle mobility problems. It comprises pushing a $30°$ apex cone of $0.5\,m^2$ base area into the ground, where the required pressure for a given penetration is given in terms of *Cone Index (CI)*, which provides a basis for trafficability.

Western button-bottom pile A pile formed by driving a concrete point in the end of a steel tube and at the required level, lowering a spirally corrugated steel shell to lock onto the point, and withdrawing the driving tube after filling the shell with concrete and reinforcement, if required. It is widely used in the north American continent. *See also* **pile foundations.**

West's Hardrive *See West's shell piling system.*

West's segmental pile *See West's shell piling system.*

West's shell piling system A system *(Figure W.2)* comprising precast concrete hollow shells which are threaded on to a steel mandrel and driven into the ground by a drop hammer, additional units being added as necessary until the required set or depth is reached. After withdrawal of the mandrel, the core is filled with concrete, which can be reinforced as required. Piles vary in size from about 380 mm diameter for working loads of 650–1200 kN and can be raked up to 1 in 3 towards or away from the piling machine. The system has the advantage of being able to design the concrete in the shell against aggressive soil or groundwater conditions, and since the core is formed after driving, it is unstressed.

West's segmental pile has an outside diameter of 280 mm and is designed for light axial loading (compression and tension) of about 300 kN using polypropylene reinforced segments of 1 m length. The segments incorporate a spigot and socket joint and are driven by a drop hammer. A central hole of 70 mm diameter allows of inspection after completion of driving and can accommodate a single reinforcement bar, which can be post-tensioned, if required, before grouting the hole.

West's Hardrive precast concrete modular piling system incorporates 285 mm × 285 mm square units of 10 m, 7.5 m, 5 m and 2.5 m length. A special Hardrive steel joint enables the individual units to be joined for severe driving conditions and can carry working loads of up to 1200 kN in compression and 200 kN in tension.

See also **pile foundations.**

wet sand process A method of stabilising granular soil by mixing with it about 6 per cent *SRO* and about 2 per cent hydrated lime.

wetness index (I_w)

$$I_w = \frac{LL - m}{LL - OMC}$$

264

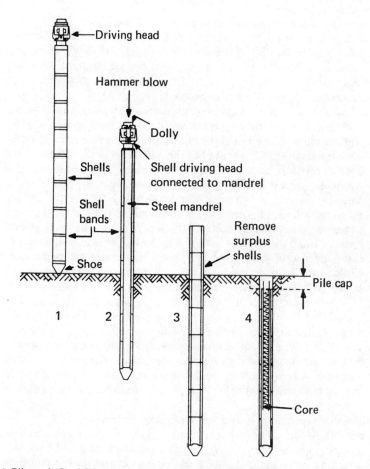

- Driving head
- Hammer blow
- Dolly
- Shell driving head connected to mandrel
- Shells
- Steel mandrel
- Shell bands
- Remove surplus shells
- Shoe
- Pile cap

1   2   3   4

- Core

1 Pile ready for driving — shells threaded on steel mandrel
2 Driving to set — blows from hammer conveyed directly to shoe
3 Inspection of pile — after removal of mandrel
4 Completed pile with compacted concrete core cast in place and reinforced as required

Figure W.2   Sequence of driving operations of West's shell pile

where LL = the *liquid limit*; m = the *moisture content*; and OMC = the *optimum moisture content* obtained in the *compaction* test.

wetting agent   An additive used to reduce the surface tension of a fluid.

whaleback fold   *See doubly plunging fold.*

wide strip foundations   *See strip foundation.*

265

wild well An *artesian well* flowing out of control.

wilting coefficient The ratio between the weight of water in the soil and the dry weight of the soil at the stage when the available moisture is insufficient to maintain healthy plant growth and the leaves start to wilt.

Wipsampler *See drillstring anchor.*

wireline drilling A system of drilling *boreholes* and *wells* using boring tools suspended on a flexible wire, as opposed to the conventional method of connecting them to a string of rigid rods. This avoids the need to break out the rods every time a core run has been completed and thus allows of more rapid progress. In the wireline system the drill string combines both *casing* and drillrods. The core barrel used to penetrate the ground has an outer and an inner section which allows the core sample taken during the core run to be recovered up to the drill deck without the necessity to remove the outer barrel from the borehole. Specially developed sampling and *in situ* testing equipment has been developed which can be run down the centre of the drill string to the bottom of the borehole, which avoids the need for extraction of the drilling equipment each time a test or sample is required. *See drillstring anchor.*

wireline drillstring *See drillstring anchor.*

wison A wireline penetrometer for carrying out *static cone penetration tests* from the bottom of a borehole. Testing is possible to depths of 450 m below ground level or sea surface. It is used with equipment similar to the Wipsampler. *See drillstring anchor; Figure D.4.*

wood-step taper piles *See Raymond piling systems.*

work In American contract law, any and all obligations, duties and responsibilities necessary to the successful completion of a project assigned to or undertaken by the contractor under the contract documents, usually including the furnishing of all labour, materials, equipment and other incidentals.
Reference: *Manual of Water Well Construction Practices*, US Environmental Protection Agency, 570/9-75-001.

working load That load which a structural member such as a pile or anchor is designed to carry.

working pile One of the piles forming the foundation of the structure.
Reference: CP 2004 : 1972.

Wulff net A circular grid of latitude and longitude projected from a sphere by *stereographic projection*. The net is uncorrected for area distortion, and so retains angular relationships from the original sphere. For this reason it is principally used as an aid to graphical calculation using the principle of stereographic projection.

Wyllie relationship   An expression for calculating *porosity* (*n*) from sonic-log times:

$$n = \frac{\Delta t - \Delta t_{ma}}{\Delta t_f - \Delta t_{ma}}$$

where Δt = the observed transit time; Δt_f = the transit time in the pore fluid; and Δt_{ma} = the transit time in the rock matrix.

X

X–Y **reader**   A device for converting the positions of points on a map or graph to digital coordinates.

Y

Yankee brob   A Z-shaped metal strap used to fix timbers together.

Young's modulus   The ratio between the normal stress in a given direction and the elastic strain in that direction. Young's moduli for rocks (*see Table 8*) are usually measured by axial compression of rock cores, or are calculated from the *seismic velocity* of the material. The Young's modulus is the slope of the load deflection curve in a compression test, and varies with the stress range considered. *See stress–strain relationships.*

yo-yo drilling   *See cable-tool drilling.*

Z

zone of aeration, or unsaturated zone   The unsaturated zone above the water table where the voids are partly filled with air. The zone may contain a *perched aquifer*.

zone of saturation   The zone below the water table where all the voids are filled with water at greater than atmospheric pressure. Also known as the *phreatic zone*.

NOTE: These values are for preliminary design purposes only, and may need alteration upwards or downwards. No addition has been made for the depth of embedment of the foundation. Reference should be made to other parts of the Code when using this Table.

Group	Class	Types of rocks and soils	Presumed bearing value		Remarks
			kN/m^2*	kgf/cm^2 or $tonf/ft^2$*	
I Rocks	1	Hard igneous and gneissic rocks in sound condition	10000	100	These values are based on the assumption that the foundations are carried down to unweathered rock
	2	Hard limestones and hard sandstones	4000	40	
	3	Schists and slates	3000	30	
	4	Hard shales, hard mudstones and soft sandstones	2000	20	
	5	Soft shales and soft mudstones	600 to 1000	6 to 10	
	6	Hard sound chalk, soft limestone	600	6	
	7	Thinly bedded limestones, sandstones, shales	To be assessed after inspection		
	8	Heavily shattered rocks			
II Non-cohesive soils	9	Compact gravel, or compact sand and gravel	>600	>6	Width of foundation (B) not less than 1 m (3 ft). Groundwater level assumed to be a depth not less than B below the base of the foundation
	10	Medium dense gravel, or medium dense sand and gravel	200 to 600	2 to 6	
	11	Loose gravel, or loose sand and gravel	<200	<2	
	12	Compact sand	>300	>3	
	13	Medium dense sand	100 to 300	1 to 3	
	14	Loose sand	<100	<1	
III Cohesive soils	15	Very stiff boulder clays and hard clays	300 to 600	3 to 6	Group III is susceptible to long-term consolidation settlement
	16	Stiff clays	150 to 300	1.5 to 3	
	17•	Firm clays	75 to 150	0.75 to 1.5	
	18	Soft clays and silts	<75	<0.75	
	19	Very soft clays and silts	Not applicable		
IV	20	Peat and organic soils			
V	21	Made ground or fill			

* 1 tonf/ft² = 1.094 kgf/cm² = 107.25 kN/m²

Table 2: STRENGTH OF COHESIVE SOILS AND ROCKS*

Term		Unconfined shear strength kN/m^2	Unconfined compressive strength MN/m^2
Soil	Rock		
Very soft	—	<20	—
Soft	—	20–40	—
Firm	—	40–75	—
Stiff	—	75–150	—
Very stiff	—	150–300	—
Hard	Very weak	300–625	0.60–1.25
—	Weak	—	1.25–5.0
—	Moderately weak	—	5.00–12.5
—	Moderately strong	—	12.5–50
—	Strong	—	50–100
—	Very strong	—	>100

*Proposed in *The Description of Rock Masses for Engineering Purposes.* Report by the Geological Society Engineering Group Working Party. Quarterly Journal *Engineering Geology*, 1977. **10**, pp. 355—388
NB: BS 5930:1981 Code of Practice for Site Investigations proposes similar strength scales for soils and rock except that it limits soils to an upper strength term of 'very stiff or hard' having an unconfined shear strength over 150 kN/m3; and rocks to a lower strength term of 'very weak' having a compressive strength under 1.25 MN/m².

Table 3: COMPRESSIBILITY OF CLAYS

Clay type	Qualitative description of compressibility	Coefficient of volume compressibility (mv) m^2/MN
Heavily overconsolidated boulder clays	Very low	<0.05
Overconsolidated boulder clays	Low	0.05–0.10
Overconsolidated clays of high plasticity	Medium	0.10–0.30
Normally consolidated alluvial clays	High	0.30–1.50
Very organic alluvial clays and peats	Very high	>1.50

NB The coefficient of volume compressibility, mv, is obtained from the pressure:voids ratio relationship in the consolidation test and its value will depend on the pressure range under consideration. For comparative purposes the mv is often quoted for a pressure range of p_0 to $p_0 + 100$ kN/m² where p_0 is the effective overburden pressure in kN/m².

269

Table 4: PLASTICITY OF SOILS

Plasticity	Range of liquid limit per cent
Low	Under 35
Medium	35–50
High	50–70
Very high	70–90
Extremely high	Over 90

Table 5: APPROXIMATE RELATIONSHIP BETWEEN STANDARD PENETRATION TEST VALUE (N) AND SHEAR STRENGTH OF COHESIVE SOILS

N-value	Consistency	Terzaghi & Peck value*	
		lb/ft^2	kN/m^2
<2	Very soft	<250	<12
2–4	Soft	250–500	12–24
4–8	Medium	500–1000	24–48
8–15	Stiff	1000–2000	48–96
15–30	Very stiff	2000–4000	96–192
>30	Hard	>4000	>192

*Soil Mechanics in Engineering Practice, K. Terzaghi and R. B. Peck, Wiley, New York, 1948.

Table 6: RELATIONSHIP BETWEEN STANDARD PENETRATION TEST VALUE (N) AND RELATIVE DENSITY OF GRANULAR SOILS

N-value	Relative density
<4	Very loose
4—10	Loose
10—30	Medium dense
30—50	Dense
>50	Very dense

270

Table 7: STRENGTH OF ROCKS

Rock type	Compressive strength	Tensile strength	Shear strength
Granite	100–250	7–25	14–50
Basalt	150–300	10–30	20–60
Sandstone	20–170	4–25	8–60
Limestone	30–250	5–25	10–50
Shale	5–100	2–10	3–30

All values in MN/m^2
Source: *Principles of Engineering Geology*, P. B. Attewell and I. W. Farmer, Chapman & Hall, 1976

Table 8: YOUNG'S MODULI OF ROCKS

Rock	E kN/mm^2
Granite	20–60
Gabbro	70–110
Basalt	60–100
Sandstone	5–80
Limestone	10–80
Shale	10–30

Based on *Engineering Properties of Rocks*, I. W. Farmer, Spon 1968

Table 9: ROCK QUALITY

Description	RQD value (%)
Very poor	0–25
Poor	25–50
Fair	50–75
Good	75–90
Excellent	90–100

271

Rock	Resistivity (ohm-metres)	
	Dry	Saturated
Granite	10^6	500–5000
Gabbro	10^5–10^7	500–5000
Shale	1000	10–1000
Limestone	10 000	100–10 000
Sandstone	1000–100 000	10–1000

Soil*	Resistivity (ohm-metres)	
	Marine	Terrestrial
Soft clay	1–10	15–50
Loose sand	1–10	15–50
Glacial till	—	500

* Soil resistivities depend on their porosity and salinity
Source: *Introduction to Geophycial Prospecting*, M. B. Dobrin, McGraw-Hill 1960 & 1976

Table 11:   SEISMIC (P-WAVE) VELOCITIES

Material	Velocity (m/s)
Granite	3000–5000
Basalt	4500–6500
Sandstone	1400–4000
Shale	1400–3000
Slate	3500–5500
Limestone	2500–6000
Clay	1100–2500
Sand	1800

Sources: *Engineering Properties of Rocks*, I. W. Farmer, Spon, 1968; *Introduction to Geophysical Prospecting*, M. B. Dobrin, McGraw-Hill 1976

Table 12:   ELASTIC CONSTANTS IN TERMS OF THE LAME CONSTANTS

Modulus of rigidity	G
Bulk modulus	$\lambda + \frac{2}{3}G$
Young's modulus	$\dfrac{G(3\lambda + 2G)}{\lambda + G}$
Poisson's ratio	$\dfrac{\lambda}{2(\lambda + G)}$

Table 13: BJERRUM'S DANGER LIMITS FOR DISTORTION OF STRUCTURES

Angular distortion	Behaviour of structure
1/750	Limit where difficulties with machinery sensitive to settlements are to be feared
1/600	Limit of danger for frames with diagonals
1/500	Safe limit for buildings where cracking is not permissible
1/300	Limit where first cracking in panel walls is to be expected
1/300	Limit where difficulties with overhead cranes are to be expected
1/250	Limit where tilting of high rigid buildings might become visible
1/150	Considerable cracking in panel walls and brick walls
1/150	Safe limit for flexible brick walls ($H/L < \frac{1}{4}$)
1/150	Limit where general structural damage of buildings is to be feared

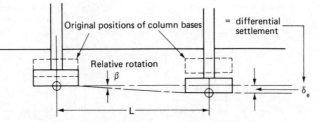

Where superstructure remains vertical:

Angular distortion = $\dfrac{\delta_e}{L}$ = relative rotation (Burland and Wroth) = β

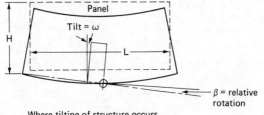

Where tilting of structure occurs

273

Table 14: USSR BUILDING CODE (1955)
Values of limiting differential settlement and deflection of building foundations

Condition	Soil	
	Sand and clay in hard condition	Clay in plastic condition
(1) Difference in settlement of column foundations for:		
(a) Steel and reinforced concrete framed structures	0.002 L	0.002 L
(b) End row columns for buildings with brick cladding	0.007 L	0.001 L
(c) Structures where auxiliary strain does not arise during non-uniform settlement of foundations	0.005 L	0.005 L
(2) Relative deflection of plain brick walls		
For L/H ≤ 3	0.0003	0.0004
For L/H ≥ 5	0.0005	0.0007

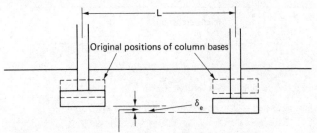

Difference in settlement = differential settlement

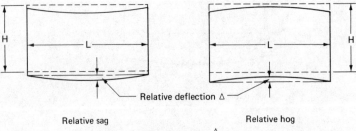

Relative deflection Δ

Relative sag   Relative hog

Deflection ratio = $\dfrac{\Delta}{L}$

274

Table 15: THE GEOLOGICAL TIMESCALE

Era	Period	Approx. age to the base (million years B.P.)
Precambrian	—	4500
Palaeozoic	Cambrian	600
	Ordovician	500
	Silurian	440
	Devonian	400
	Carboniferous	350
	Permian	280
Mesozoic	Triassic	230
	Jurassic	200
	Cretaceous	140
Cainozoic	Palaeogene Tertiary	70
	Neogene Quaternary	2

Table 16: DETAILS OF DRILLING EQUIPMENT FOR ROTARY DRILLING RIGS

Name	Purpose	Sizes*
Swivel	Connected to top of drill rods to provide watertight rotating joint allowing flushing media under pressure to be pumped to bottom of hole while drill tools are rotating	Sizes according to drill string diameter on light and heavy duty
Rods	To connect core barrel or bit to the drill	XRT (28 mm) through to HW (89 mm)
Core barrel	For taking rock core sample. Various types available— single and double tube Q-series wire line, etc.	Giving core sizes from 17 mm to 165 mm diameter
Core bit	Placed at bottom of core barrel and has annulus set with diamond or hard material, e.g. Stellite	Various to suit size of core required
Rock roller bit	Two or three rotating cones enable open boring (i.e. no core recovered)	Various
Casing	To enable penetration through loose or collapsed ground	XRT (36.5 mm) through to ZX (219 mm)
Reaming shell	Placed between bit and barrel— reams out hole and protects barrel from excessive wear	Various to suit bits
Subs (adaptors)	To join different tools to different strings or to adapt from one string of rods to another	Various to suit rods, barrels, etc.
Core catcher	To catch core as it enters core barrel—by wedging action	Various to suit core barrels
Foot clamps	To hold rods when lowering	To fit rod size
Water pump or air compressor	To provide flushing medium to clear chippings away from bit and to cool the bit	To suit hole size
Core boxes	To take recovered core and preserve it	To suit core size

* See Table 17 for more exact information on sizes

Table 17: BS 4019 GENERAL PARTICULARS OF RODS, CASING AND CORE BARRELS
(dimensions in inches, metric dimensions in parentheses)

Drill rod	Rod O.D.	Rod coupling I.D.	Casing	Casing O.D.	Casing coupling I.D. min	Core barrel		Core bit set O.D.	Core bit set I.D.	Core barrel reaming shell O.D.	Hole Dia. nom.	Core Dia. nom.
XRT	1.094 (27.79)	0.406 (10.31)	XRT	1.437 (36.50)	1.190 (30.23)	XRT		1.160 (29.46)	0.735 (18.67)	1.175 (29.84)	$1\frac{3}{16}$ (30.2)	$\frac{11}{16}$ (17.5)
EW	1.375 (34.92)	0.437 (11.11)	EX	1.812 (46.02)	1.500 (38.10)	EWX	EWM	1.470 (37.34)	0.845 (21.46)	1.485 (37.72)	$1\frac{1}{2}$ (38.1)	$\frac{13}{16}$ (20.6)
AW	1.750 (44.45)	0.625 (15.88)	AX	2.250 (57.15)	1.906 (48.41)	AWX	AWM	1.875 (47.62)	1.185 (30.10)	1.890 (48.01)	$1\frac{15}{16}$ (49.2)	$1\frac{3}{16}$ (30.2)
BW	2.125 (53.98)	0.750 (19.05)	BX	2.875 (73.02)	2.375 (60.32)	BWX	BWM	2.345 (59.56)	1.655 (42.04)	2.360 (59.94)	$2\frac{3}{8}$ (60.3)	$1\frac{5}{8}$ (41.3)
NW	2.625 (66.68)	1.375 (34.92)	NX	3.500 (88.90)	3.000 (76.20)	NWX	NWM	2.965 (75.31)	2.155 (54.74)	2.980 (75.69)	3 (76.2)	$2\frac{1}{8}$ (54.0)
HW	3.500 (88.90)	2.375 (60.32)	HX	4.500 (114.30)	3.937 (100.00)	HWX	HWF	3.890 (98.81)	3.000 (76.20)	3.906 (99.21)	$3\frac{15}{16}$ (100.0)	3 (76.2)
			PX	5.500 (139.70)	4.875 (123.82)		PF	4.720 (119.89)	3.625 (92.08)	4.750 (120.65)	$4\frac{3}{4}$ (120.6)	$3\frac{5}{8}$ (92.1)
			SX	6.625 (168.28)	5.875 (149.23)		SF	5.720 (145.29)	4.437 (112.71)	5.750 (146.05)	$5\frac{3}{4}$ (146.0)	$4\frac{7}{16}$ (112.7)
			UX	7.625 (193.68)	6.937 (176.20)		UF	6.845 (173.86)	5.500 (139.70)	6.875 (174.62)	$6\frac{7}{8}$ (174.6)	$5\frac{1}{2}$ (139.7)
			ZX	8.625 (219.08)	7.937 (201.60)		ZF	7.845 (199.26)	6.500 (165.10)	7.875 (200.02)	$7\frac{7}{8}$ (200.0)	$6\frac{1}{2}$ (165.1)

Table 18:   MODIFIED MERCALLI INTENSITY (DAMAGE) SCALE OF 1931 (ABRIDGED)

I   Not felt except by a very few under especially favourable circumstances.

II   Felt only by a few persons at rest, especially on upper floors of buildings. Delicately suspended objects may swing.

III   Felt quite noticeably indoors, especially on upper floors of buildings, but many people do not recognise it as an earthquake. Standing motorcars may rock slightly. Vibration like passing truck. Duration estimated.

IV   During the day felt indoors by many, outdoors by few. At night some awakened. Dishes, windows, and doors disturbed; walls make creaking sound. Sensation like heavy truck striking building. Standing motorcars rocked noticeably.

V   Felt by nearly everyone; many awakened. Some dishes, windows, etc. broken. A few instances of cracked plaster. Unstable objects overturned. Disturbances of trees, poles, and other tall objects sometimes noticed. Pendulum clocks may stop.

VI   Felt by all; many frightened and run outdoors. Some heavy furniture moved. A few instances of fallen plaster or damaged chimneys. Damage slight.

VII   Everybody runs outdoors. Damage negligible in buildings of good design and construction; slight to moderate in well-built ordinary structures; considerable in poorly-built or badly-designed structures. Some chimneys broken. Noticed by persons driving motorcars.

VIII   Damage slight in specially-designed structures; considerable in ordinary substantial buildings, with partial collapse; great in poorly-built structures. Panel walls thrown out of frame structures. Fall of chimneys, factory stacks, columns, monuments, walls. Heavy furniture overturned. Sand and mud ejected in small amounts. Changes in well water. Persons driving motorcars disturbed.

IX   Damage considerable in specially-designed structures; well-designed frame structures thrown out of plumb; great in substantial buildings, with partial collapse. Buildings shifted off foundations. Ground cracked conspicuously. Underground pipes broken.

X   Some well-built wooden structures destroyed; most masonry and frame structures destroyed with foundations. Ground badly cracked. Rails bent. Landslides considerable from river banks and steep slopes. Shifted sand and mud. Water splashed (slopped) over banks.

XI   Few, if any (masonry), structures remain standing. Bridges destroyed. Broad fissures in ground. Underground pipelines completely out of service. Earth slumps and land slips in soft ground. Rails bent greatly.

XII   Damage total. Waves seen on ground surfaces. Lines of sight and level distorted. Objects thrown upward into the air.

Table 19: JAPAN METEOROLOGICAL AGENCY INTENSITY SCALE

O *No sensation:* registered by seismographs but no perception by the human body.

I *Slight:* felt by persons at rest or persons especially sensitive to earthquakes.

II *Weak:* felt by most persons. Slight rattling of doors and Japanese latticed paper sliding doors (shōji).

III *Rather strong:* shaking of houses and buildings. Heavy rattling of doors and shōji, swinging of chandeliers and other hanging objects. Movement of liquids in vessels.

IV *Strong:* strong shaking of houses and buildings. Overturning of unstable objects. Spilling of liquids out of vessels four-fifths full.

V *Very strong:* cracking of plaster walls. Overturning of tombstones and stone lanterns. Damage to masonry chimneys and mud-plastered warehouses.

VI *Disastrous:* demolition of up to 30 per cent of Japanese wooden houses. Numerous landslides and embankment failures; fissures on flat ground.

VII *Ruinous:* demolition of more than 30 per cent of Japanese wooden houses.

Table 20: CONVERSION TABLES

Area

km^2	ha	m^2	mm^2	$sq\ mile$	$acre$	yd^2	ft^2	in^2
1	10^2	10^6	10^{12}	0.3861	247.11	1.196×10^6	1.076×10^7	1.550×10^9
10^{-2}	1	10^4	10^{10}	3.861×10^{-3}	2.4711	1.196×10^4	1.076×10^5	1.550×10^7
10^{-6}	10^{-4}	1	10^6	3.861×10^{-7}	2.471×10^{-4}	1.1960	10.764	1.550×10^3
10^{-12}	10^{-10}	10^{-6}	1	3.861×10^{-13}	2.471×10^{-10}	1.196×10^{-6}	1.076×10^{-5}	1.550×10^{-3}
2.590	2.590×10^2	2.590×10^6	2.590×10^{12}	1	640	3.098×10^6	2.788×10^7	4.014×10^9
4.047×10^{-3}	0.4047	4047	4.047×10^9	1.563×10^{-3}	1	4840	43560	6.273×10^6
8.361×10^{-7}	8.361×10^{-5}	0.8361	8.361×10^5	3.228×10^{-7}	2.066×10^{-4}	1	9	1296
9.290×10^{-8}	9.290×10^{-6}	9.290×10^{-2}	9.290×10^4	3.587×10^{-8}	2.296×10^{-5}	0.111	1	144
6.452×10^{-10}	6.452×10^{-8}	6.452×10^{-4}	6.452×10^2	2.491×10^{-10}	1.594×10^{-7}	7.716×10^{-4}	6.944×10^{-3}	1

Permeability

m/s	cm/s	$m/year$	$Darcy$	ft/day	$ft/year$	ft/min	$lugeon*$
1	10^2	3.156×10^7	1.035×10^5	2.835×10^5	1.035×10^8	196.85	1×10^7
10^{-2}	1	3.156×10^5	1.035×10^3	2.835×10^3	1.035×10^6	1.968	1×10^5
3.169×10^{-8}	3.169×10^{-6}	1	3.281×10^{-3}	8.982×10^{-3}	3.281	6.238×10^{-6}	0.317
9.659×10^{-6}	9.659×10^{-4}	304.8	1	2.738	10^3	1.901×10^{-3}	96.57
9.659×10^{-9}	9.659×10^{-7}	0.3048	10^{-3}	2.738×10^{-3}	1	1.901×10^{-6}	9.66×10^{-2}
3.528×10^{-6}	3.528×10^{-4}	111.33	0.365	1	365.25	6.944×10^{-4}	35.27
5.080×10^{-3}	0.508	1.603×10^5	5.260×10^2	1.440×10^3	5.260×10^5	1	5.08×10^4
1×10^{-7}	1×10^{-5}	3.156	1.035×10^{-2}	2.835×10^{-2}	10.35	1.968×10^{-5}	1

* Defined by Gignoux & Barbeir, *Géologie des Barrages* 1955, p. 281 as the water acceptance of 1 litre/metre/minute in a BX (60.3 mm) borehole at a pressure of 10 kg/cm² (approx 10 atmospheres) over a period of 10 minutes and is *roughly* equivalent to a permeability of 10^{-5} cm/sec

Length

km	m	mm	mile	yard	ft	in
1	10^3	10^6	0.6214	1.094×10^3	3.281×10^3	3.937×10^4
10^{-3}	1	10^3	6.214×10^{-4}	1.0936	3.281	39.370
10^{-6}	10^{-3}	1	6.214×10^{-7}	1.094×10^{-3}	3.281×10^{-3}	3.937×10^{-2}
1.6093	1.609×10^3	1.609×10^6	1	1760	5280	63360
9.144×10^{-4}	0.9144	9.144×10^2	5.682×10^{-4}	1	3	36
3.048×10^{-4}	0.3048	3.048×10^2	1.894×10^{-4}	0.3333	1	12
2.54×10^{-5}	2.54×10^{-2}	25.4	1.578×10^{-5}	2.778×10^{-2}	8.333×10^{-2}	1

Pressure, stress and modulus of elasticity

N/mm^2 MN/m^2 MPa	kN/m^2 kPa	kp/cm^2 kgf/cm^2	bar	atm^*	$m\,H_2O$	$ft\,H_2O$	$mm\,Hg$	$tonf/ft^2$	psi lbf/in^2	lbf/ft^2
1	10^3	10.197	10.000	9.869	101.97	334.55	7500.6	9.324	145.04	20885
10×10^{-3}	1	1.020×10^{-2}	10^{-2}	9.869×10^{-3}	0.1020	0.3346	7.5006	9.324×10^{-3}	0.14504	20.885
9.807×10^{-2}	98.07	1	0.9807	0.9678	10.000	32.808	735.56	0.9144	14.223	2048.17
0.100	10^2	1.0197	1	0.9869	10.197	33.455	750.06	0.9324	14.504	2088.5
0.101325	101.325	1.0332	1.01325	1	10.332	33.899	760.00	0.9447	14.696	2116.2
9.8067×10^{-3}	9.8067	10.000×10^{-2}	9.8067×10^{-2}	9.678×10^{-2}	1	3.2808	73.556	9.144×10^{-2}	1.4223	204.82
2.989×10^{-3}	2.989	3.048×10^{-2}	2.989×10^{-2}	2.945×10^{-2}	0.3048	1	22.420	2.781×10^{-2}	0.43275	62.316
1.333×10^{-4}	0.1333	1.3595×10^{-3}	1.333×10^{-3}	1.316×10^{-3}	1.360×10^{-2}	4.460×10^{-2}	1	1.243×10^{-3}	1.934×10^{-2}	2.7845
0.10725	107.252	1.0937	1.0725	1.0588	10.937	35.881	804.46	1	15.556	2240
6.895×10^{-3}	6.895	7.031×10^{-2}	6.895×10^{-2}	6.805×10^{-2}	0.7031	2.3067	51.715	6.429×10^{-2}	1	144
4.788×10^{-5}	4.788×10^{-2}	4.882×10^{-4}	4.788×10^{-4}	4.725×10^{-4}	4.882×10^{-3}	1.602×10^{-2}	0.3591	4.464×10^{-4}	6.944×10^{-3}	1

* Standard Physical Atmosphere = 1013.25 mbar at temperature of 15°C and dry

Volume

m^3	litre	cm^3, ml	yd^3	ft^3	in^3	UK gallon	US gallon
1	10^3	10^6	1.308	35.315	61024	219.97	264.17
10^{-3}	1	10^3	1.308×10^{-3}	3.531×10^{-2}	61.024	0.2200	0.2642
10^{-6}	10^{-3}	1	1.308×10^{-6}	3.531×10^{-5}	6.102×10^{-2}	2.199×10^{-4}	2.642×10^{-4}
0.7646	764.6	7.646×10^5	1	27	46656	168.19	201.99
2.832×10^{-2}	28.317	2.832×10^4	3.704×10^{-2}	1	1728	6.229	7.481
1.639×10^{-5}	1.639×10^{-2}	16.387	2.143×10^{-5}	5.787×10^{-4}	1	3.605×10^{-3}	4.329×10^{-3}
4.546×10^{-3}	4.546	4.546×10^3	5.946×10^{-3}	0.1605	277.42	1	1.201
3.785×10^{-3}	3.785	3.785×10^3	4.951×10^{-3}	0.1337	231.00	0.8327	1

Mass

Tonne, Mg	kg	g	UK ton	US ton	lb	oz
1	10^3	10^6	0.9842	1.1023	2.205×10^3	3.527×10^4
10^{-3}	1	10^3	9.842×10^{-4}	1.102×10^{-3}	2.20462	35.274
10^{-6}	10^{-3}	1	9.842×10^{-7}	1.102×10^{-6}	2.205×10^{-3}	3.527×10^{-2}
1.016	1016.1	1.016×10^6	1	1.12	2240	35840
0.9072	907.2	9.072×10^5	0.8929	1	2000	32000
4.536×10^{-4}	0.4536	453.6	4.464×10^{-4}	5.000×10^{-4}	1	16
2.835×10^{-5}	2.835×10^{-2}	28.3495	2.790×10^{-5}	3.125×10^{-5}	6.25×10^{-2}	1

Force and weight

MN	kN	N	tonne	kgf	ton f	lb f
1	10^3	10^6	1.0197×10^2	1.0197×10^5	100.4	2.248×10^5
10^{-3}	1	10^3	0.10197	1.0197×10^2	0.1004	224.81
10^{-6}	10^{-3}	1	1.0197×10^{-4}	0.10197	1.004×10^{-4}	0.2248
9.807×10^{-3}	9.807	9.807×10^3	1	10^3	9.842×10^{-1}	2.2046×10^3
9.807×10^{-6}	9.807×10^{-3}	9.807	10^{-3}	1	9.842×10^{-4}	2.2046
9.964×10^{-3}	9.96402	9.964×10^3	1.01605	1016.05	1	2240
4.448×10^{-6}	4.448×10^{-3}	4.44822	4.536×10^{-4}	0.45359	4.464×10^{-4}	1

Density

Tonne/m^3 Mg/m^3 g/cm^3	kg/m^3	lb/in^3	UK ton/yd^3	US ton/yd^3	lb/ft^3
1	10^3	0.03613	0.75248	0.8428	62.423
10^{-3}	1	3.613×10^{-5}	7.525×10^{-4}	8.428×10^{-4}	6.243×10^{-2}
27.680	2.768×10^4	1	20.828	23.328	1.728×10^3
1.3289	1.329×10^3	4.801×10^{-2}	1	1.120	82.963
1.1866	1.187×10^3	4.287×10^{-2}	0.8929	1	74.074
1.602×10^{-2}	16.019	5.787×10^{-4}	1.205×10^{-2}	1.350×10^{-2}	1